SpringerBriefs in Modern Perspectives on Disability Research

Series Editors

Gabriel Bennett, Independent Researcher, Klemzig, Australia

Emma Goodall, Healthy Possibilities, Seaford, Australia

This book series on disability research is a comprehensive collection of research on disability and related issues. The series is designed to promote interdisciplinary collaboration and exchange, bringing together scholars and practitioners from different fields to share their perspectives and insights. Disability research is an interdisciplinary field that examines the social, cultural, historical, and political dimensions of disability. It encompasses a wide range of topics, including disability rights, accessibility, assistive technologies, healthcare, education, employment, and social welfare. Disability research scholars employ a range of theoretical and methodological approaches to understand the experiences of people with disabilities, as well as the ways in which disability intersects with other social identities such as race, gender, sexuality, and class.

The series seeks to advance knowledge and understanding of disability by publishing rigorous, innovative, and relevant research. It aims to promote disability rights and social justice by highlighting the ways in which people with disabilities are marginalized and discriminated against in society, and advocating for greater social inclusion and accessibility. The series also seeks to inform policy and practice by disseminating research findings that can help to shape policy decisions and contribute to positive social change.

Gabriel Bennett

Senotherapy

A Potential Pharmacological Strategy
for Prolonging Human Lifespan
and Healthspan

 Springer

Gabriel Bennett
Independent Researcher
Klemzig, SA, Australia

ISSN 3004-9709   ISSN 3004-9717   (electronic)
SpringerBriefs in Modern Perspectives on Disability Research
ISBN 978-981-97-3636-2   ISBN 978-981-97-3637-9   (eBook)
https://doi.org/10.1007/978-981-97-3637-9

This Springer imprint is published by the registered company Springer Nature Singapore Pte Ltd.
The registered company address is: 152 Beach Road, #21-01/04 Gateway East, Singapore 189721,
Singapore

Statement About the Usage of Artificial Intelligence

ChatGPT, a large language model artificial intelligence program developed by the company OpenAI, was used to enhance the clarity of the concepts articulated in this book. All other procedures that were used to create this research were performed by Dr Gabriel Bennett.

Contents

About the Author

Gabriel Bennett, PhD Dr. Gabriel Bennett, the pen name for Dr. Matthew Bennett, holds a PhD in Disability Studies from Flinders University, Australia. He has lectured in Disability Studies at Griffith University, Queensland. He has also advised the Australian Government's AutismCRC and has published articles for the *Journal of Autism and Developmental Disorders*. He is actively involved in supporting autistics and exploring key issues concerning disability. He has published several books, including:

Bennett, M., Webster, A. A., Goodall, E., & Rowland, S. (2019). *Life on the autism spectrum: Translating myths and misconceptions into positive futures*. Springer Nature. https://doi.org/10.1007/978-981-13-3359-0

Bennett, M., & Goodall, E. (2021). *Employment of persons with autism: A scoping review*. Springer Nature. https://doi.org/10.1007/978-3-030-82174-6

Bennett, M., & Goodall, E. (2022). *Addressing underserved populations in autism spectrum research: An intersectional approach*. Emerald Publishing Group. https://doi.org/10.1108/9781803824635

Bennett, G. (2023). *Autistic people in dental and medical clinics: Challenges and solutions*. Springer Nature. https://doi.org/10.1007/978-981-99-2359-5

Abbreviations

CKD	Chronic kidney disease
D + Q	Dasatinib and quercetin
DEN	Diethylnitrosamine
HCC	Hepatocellular carcinoma
HFD	High fat diet
ICD-11	International Classification of Diseases 11th Revision
IPF	Idiopathic pulmonary fibrosis
LGBT	Lesbian, Gay, Bisexual, and Transgender
MILES	Metformin in longevity study
NAFLD	Non-alcoholic fatty liver disease
NASH	Non-alcoholic steatohepatitis
SASP	Senescence-associated secretory phenotype
TAME	Targeting Ageing with Metformin
WHO	World Health Organization

List of Figures

Chapter 1
Introduction to Senotherapies

Abstract Traditionally, biological ageing was regarded as a process that could not be altered and prevented. However, with a greater proportion of the human population becoming elderly, ageing itself is starting to be considered worthy of being slowed or possibly halted. Along with lifestyle changes and other medical interventions, senotherapeutics have the potential to increase human lifespan and healthspan by reducing the accumulation of senescent cells, which can slow the ageing process. To understand their potential, this chapter begins by furnishing the reader with an overview of our current understanding of the mechanisms and hallmarks of ageing. It then explains the benefits that optimal levels of senescent cells provide, including wound healing and tumor suppression, and the detrimental impacts of senescent cell accumulation beyond an optimal level. This chapter then presents an overview of senotherapy, followed by an explanation for why this book fits the SpringerBrief series about disability, the intended audience for this book, this book's pedagogical features, and a description of the upcoming chapters.

Keywords Biological ageing · Hallmarks of biological ageing · Senescence-associated secretory phenotype (SASP) · Senescent cell accumulation · Senescent cells · Senolytics · Senomorphics · Senotherapy

1.1 Hallmarks of Biological Ageing

Humanity's achievements in extending human lifespan can be attributed to societal changes (e.g., improved sanitation), lifestyle changes (e.g., better quality of food and opportunities to exercise), and preventative medicines (e.g., vaccinations). However, these achievements have not improved human healthspan, which is the amount of time an individual lives without age-associated diseases (Borghesan et al., 2020). Consequently, in societies where the population's lifespan has increased, there has also been an increase in the number of elderly people with age-associated diseases and disabilities, such as Alzheimer's disease (Schmitt, 2017; Turrini et al., 2023). The geroscience hypothesis suggests that since biological

© The Author(s), under exclusive license to Springer Nature Singapore Pte Ltd. 2024
G. Bennett, *Senotherapy*, SpringerBriefs in Modern Perspectives on Disability Research, https://doi.org/10.1007/978-981-97-3637-9_1

ageing is the underlying cause for most age-associated diseases and disabilities, interventions that can slow or reverse biological ageing would also prevent, delay, or alleviate multiple age-associated diseases and disabilities (Pignolo et al., 2020). Thus, classifying biological ageing as a disease itself, that requires treatment, is being debated as a feasible strategy to prevent or treat many age-associated diseases and disabilities (Bulterijs et al., 2015).

There are several distinct indicators (i.e., hallmarks) of biological ageing. López-Otín et al. (2013) identified nine common hallmarks, which were (1) genomic instability, (2) shortened telomeres, (3) changes in epigenetics, (4) decline in protein regulation, (5) disruption of nutrient sensing, (6) mitochondrial dysfunction, (7) cellular senescence, (8) depletion of stem cells, and (9) altered communication between cells (Guo et al., 2022; López-Otín et al., 2013). During 2022, Schmauck-Medina and colleagues added five extra hallmarks to López-Otín et al.'s list of nine hallmarks for ageing. These five extra hallmarks were (1) compromised autophagy, (2) microbiome disturbance, (3) altered mechanical properties, (4) splicing dysregulation, and (5) inflammation (Schmauck-Medina et al., 2022) (see Fig. 1.1).

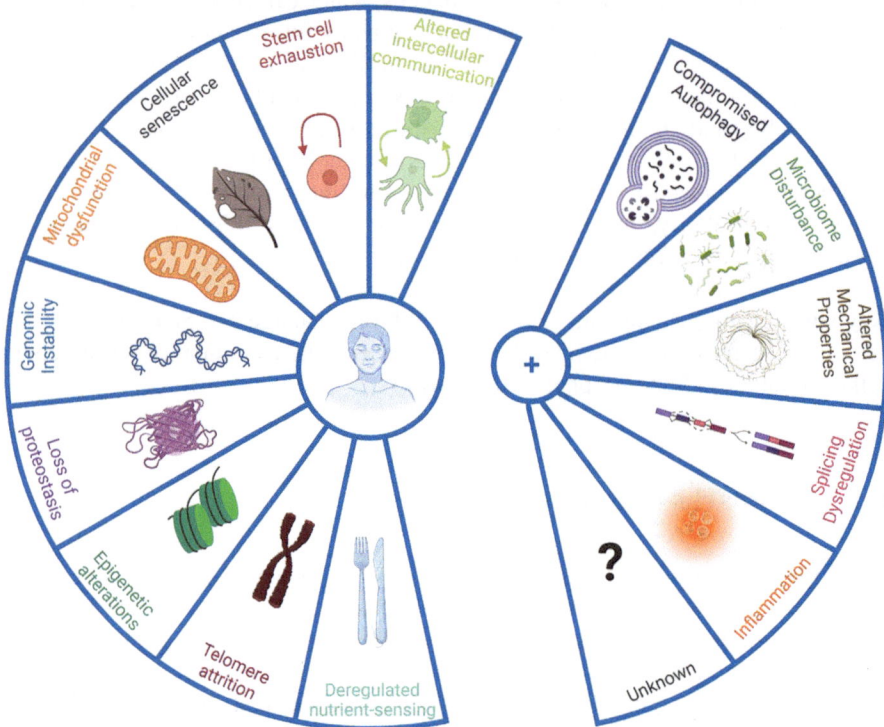

Fig. 1.1 Hallmarks of biological ageing according to Schmauck-Medina and colleagues. Notes: To qualify as a hallmark of human ageing, the processes should change with biological age not simply in a correlative manner but have a causal role. Hence, interventions that address the hallmarks should, at the very least, halt further detrimental aspects of ageing and preferably improve phenotypes associated with ageing. (Source: Schmauck-Medina et al., 2022, p. 6831)

1.2 Senescent Cells and Age-Associated Diseases

It is beyond the scope of this book to explain each hallmark of biological ageing that Schmauck-Medina et al. (2022) and López-Otín et al. (2013) have outlined. Instead, it focuses on the role that senescent cell accumulation has on the development of age-associated diseases that occur among the elderly.

1.2.1 Senescent Cell Accumulation

As people biologically age, they accumulate senescent cells (i.e., cells that have stopped dividing but have not undergone pre-programmed cell death, otherwise known as *apoptosis*) (step 1). Senescent cells emit senescence-associated secretory phenotype (SASP) molecules that aid with wound repair and tumor suppression. SASPs also alert the immune system that the senescent cells emitting the SASPs should be destroyed (step 2). Once these senescent cells have been cleared (step 3), healthy cells divide to regenerate the organ and restore optimal organ functioning (step 4). However, as biological age progresses, senescent cells accumulate in the immune system itself which reduces the immune system's ability to destroy senescent cells (step 5). This reduced ability results in senescent cells not being cleared, and neighboring cells being constantly exposed to SASPs which, in time, also make them senescent cells. The accumulation of senescent cells contributes to the creation of many age-associated diseases, such as cancer, osteoarthritis, and type 2 diabetes (step 6) (Benhamú et al., 2022; Borghesan et al., 2020; von Kobbe, 2019; Zhang et al., 2022) (see Fig. 1.2).

1.2.2 Increased Amounts of SASPs Contribute to Senescent Cell Development

At young biological ages, most people have an optimal quantity of senescent cells that secrete transient levels of SASPs. These levels of SASPs are beneficial as they produce anti-inflammatory effects, help cells that are part of the immune system identify and destroy senescent cells, and assist with tumor suppression, wound repair, and organ regeneration. However, as biological age progresses, the volume of senescent cells within the immune system increases, which compromises its ability to clear senescent cells and prevent the SASPs that they release. The persistent release of SASPs helps create more senescent cells. Once the accumulation of senescent cells exceeds a safe therapeutic threshold, organ damage occurs along with the development of age-associated diseases, such as type 2 diabetes (Benhamú et al., 2022; Borghesan et al., 2020; Chaib et al., 2022; Zhang et al., 2022) (see Fig. 1.3).

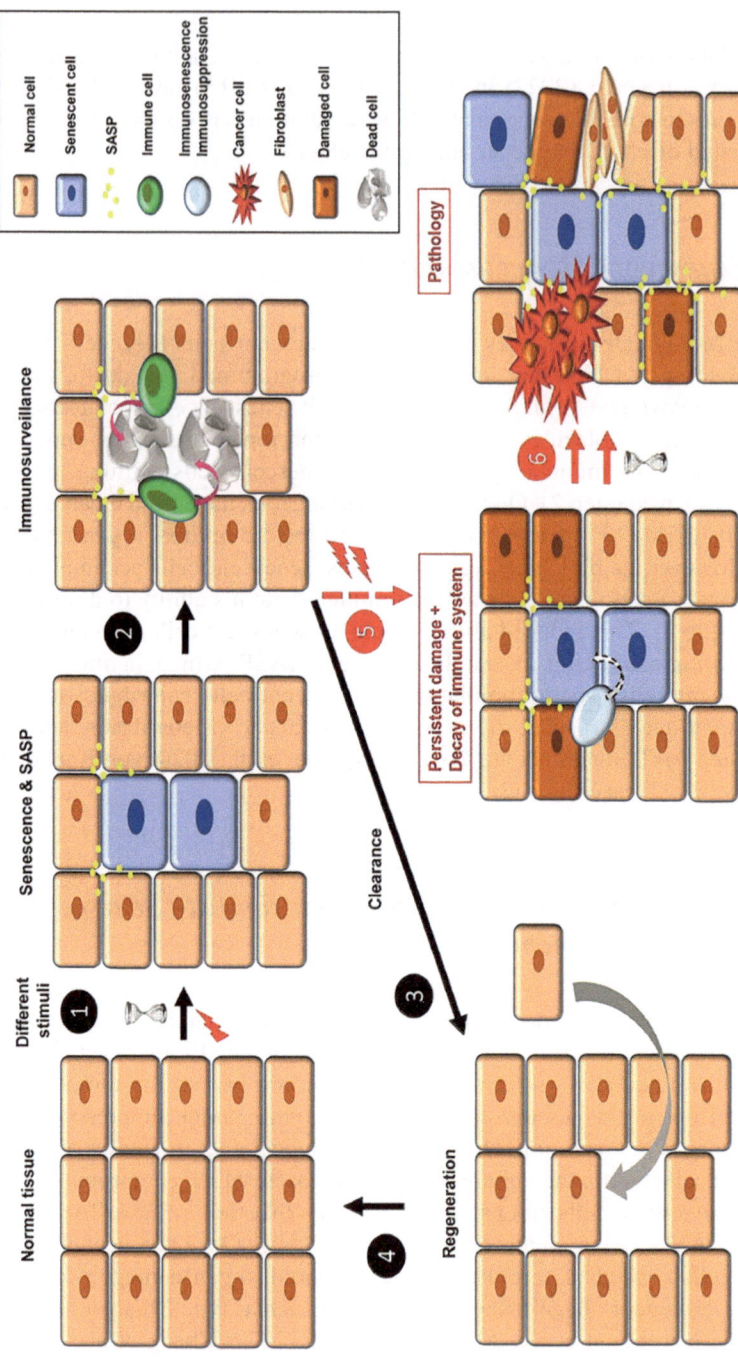

Fig. 1.2 Process of developing cellular senescence. (Source: von Kobbe, 2019, p. 12846)

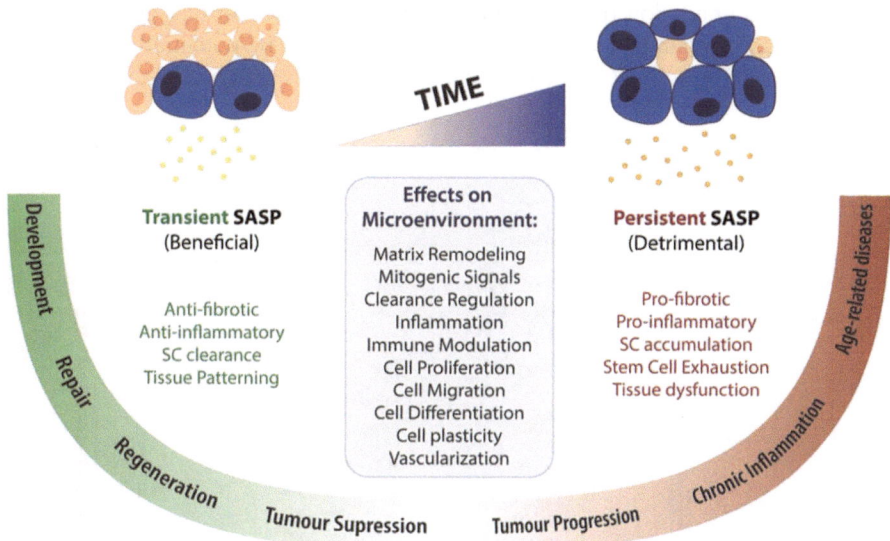

Fig. 1.3 Schematic representation of SASP-mediated biological activity in cell senescence. Note: Long-term chronic SASP is closely related to the occurrence of disease. (Source: Zhang et al., 2022, p. 2)

1.3   Senotherapeutics: Treating Senescent Cells

1.3.1   An Introduction to Senotherapeutics

There are several scientifically examined interventions that can increase human lifespan and healthspan, such as calorie restriction, gene therapy, and exercise (Guo et al., 2022). It is outside the scope of this book to summarize the literature about each intervention. Instead, the emerging field of senotherapeutics will be critiqued.

As explained previously, the accumulation of senescent cells is considered the primary cause of many age-associated diseases. Thus, by reducing this accumulation to an optimum level, it is possible that the age-associated diseases that they contribute to developing would be treated (see *geroscience hypothesis* in the glossary). The emerging field of medical research and pharmaceuticals aimed at either removing senescent cells or mitigating their detrimental effects is called *senotherapeutics* or *senotherapy* (Benhamú et al., 2022).

There are two main types of senotherapeutic medications that are being designed and tested to delay or prevent the onset of age-associated diseases. First, senolytic medications are medications that eliminate senescent cells. Second, unlike senolytics, senomorphic medications thwart the ability of senescent cells to make other cells senescent via continual exposure to SASPs. The goal of senotherapeutics is to reduce the accumulation of senescent cells to a safe level so that age-associated diseases caused by an abundance of senescent cells can be potentially avoided or eliminated (Chaib et al., 2022; Borghesan et al., 2020) (see Figs. 1.4 and 1.5).

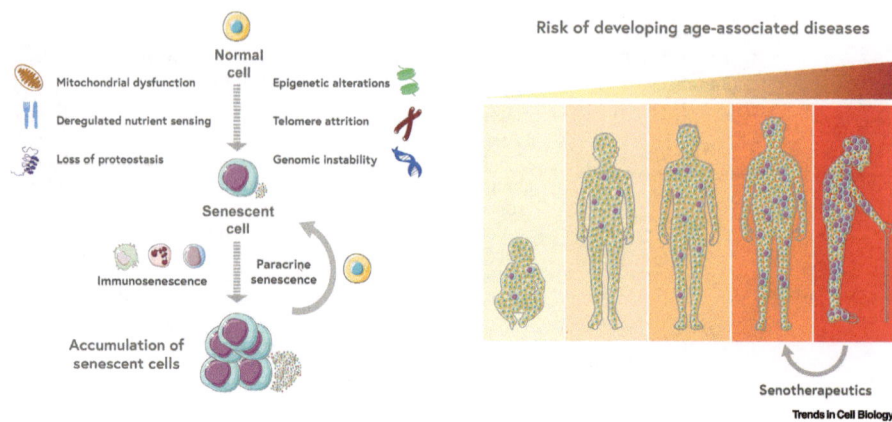

Fig. 1.4 Senescence-centric view of ageing. Notes: Some of the hallmarks of ageing (mitochondrial dysfunction, deregulated nutrient sensing, loss of proteostasis, epigenetic alterations, telomere attrition, and genomic instability) induce normal cells to become senescent, which in turn can induce paracrine senescence in nearby normal cells through senescence-associated secretory phenotype (SASP). Senescence promotion through SASP together with a decline in the immune system activity converges to induce organismal accumulation of senescent cells. In aged individuals, chronic accumulation of senescent cells contributes to tissue dysfunction and increased risk of age-associated disease development. Theoretically, senescent cell elimination with different senotherapeutic medications can improve healthspan and lifespan in aged individuals. (Source: Borghesan et al., 2020, p. 779)

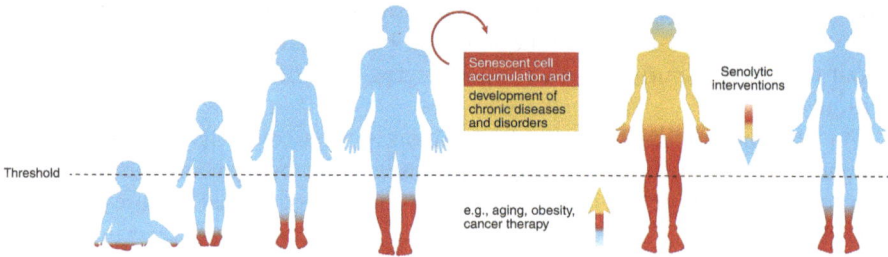

Fig. 1.5 The threshold theory of senescent cell accumulation. Notes: The threshold theory of senescent cell accumulation postulates that once senescent cell burden exceeds a threshold, self-amplifying paracrine and endocrine spread of senescence through the SASP outpaces clearance of senescent cells by the immune system. Additionally, increased abundance of SASP factors may impede immune system function, further amplifying accumulation of senescent cells. Senescent cell accumulation may also accelerate other fundamental ageing mechanisms. Hence, senescent cells with a proapoptotic, inflammatory SASP may need to exceed a threshold to exert detrimental effects. Systemic clearance of senescent cells by genetic or pharmacologic means tends to attenuate the other pillars of ageing and can delay, prevent, or alleviate multiple age-associated disorders and diseases. (Source: Chaib et al., 2022, p. 1559)

1.3.2   Senolytics

Chaib et al. (2022) have proposed that the development of age-associated diseases occurs once the proportion of senescent cells exceeds a safe threshold. Senolytic medications have the potential to reduce the accumulation of senescent cells below this threshold. Theoretically, this reduction could reverse the progression or initial development of age-associated diseases.

von Kobbe (2019) has illustrated how senolytic medications can assist immune cells with the clearance of senescent cells. When cells become senescent, they emit SASPs that damage neighboring cells (step 1). Senolytic medications assist the immune system by destroying senescent cells (step 2). The clearance of senescent cells results in the regeneration of the organ (step 3). Failure to use senolytic medications can result in the proliferation of senescent cells and aid in the development of age-associated diseases (step 4) (see Fig. 1.6) (von Kobbe, 2019).

1.3.3   Senomorphics

Unlike senolytic medications that destroy senescent cells, senomorphics are medications that prevent senescent cells from emitting SASP. von Kobbe (2019) has also illustrated how senomorphics can prevent the development of senescent cells. When senescent cells develop, they release SASPs that senomorphics target and neutralize (step 1). Once SASPs have been minimized, the body's immune cells then destroy the senescent cells (step 2) which allows for the organ to regenerate itself (step 3). If senomorphics are not provided, then the senescent cells accumulate, resulting in the progression of age-associated diseases (step 4) (von Kobbe, 2019) (see Fig. 1.7).

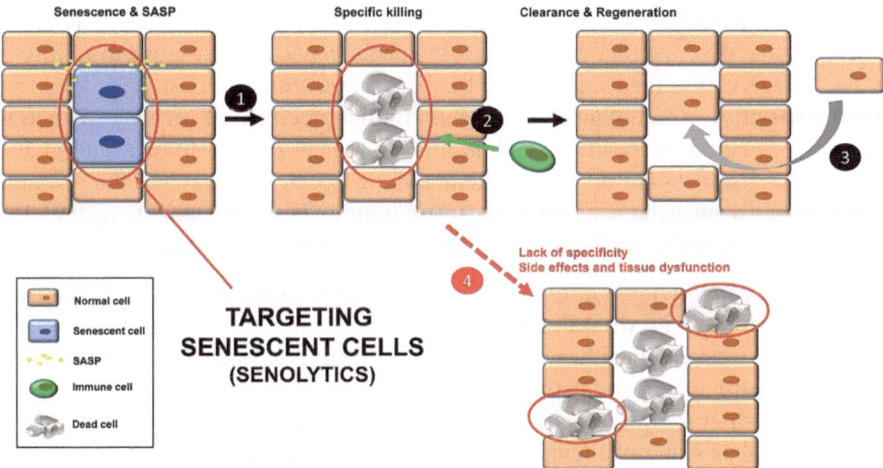

Fig. 1.6 Treatment with senolytics. (Source: von Kobbe, 2019, p. 12848)

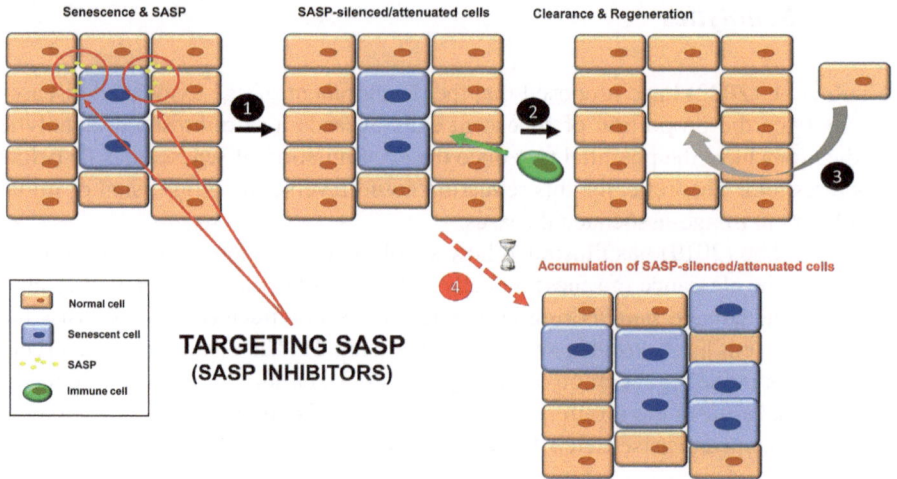

Fig. 1.7 Treatment with senomorphics to inhibit SASP factors in senescent cells. (Source: von Kobbe, 2019, p. 12848)

1.4   Relevance of This Book to the *Modern Perspectives on Disability Research* SpringerBrief Series

Occasionally, what is considered a disability changes. For example, in 1973 the American Psychiatric Association eliminated homosexuality from the second edition of its *Diagnostic and Statistical Manual*. Consequently, ending the notion that homosexuality was a mental disorder resulted in several societal changes in different countries, including implementation of legislation safeguarding the human rights of lesbian, gay, bisexual, and transgender (LGBT) individuals both in society and the workplace and allowing LGBT individuals to openly serve in the military (Bulterijs et al., 2015).

Books that describe fundamental developments in what constitutes a disability or a disease are within the scope of the *Modern Perspectives on Disability Research* SpringerBrief series. Humanity is currently on the precipice of revising the notion of ageing as being an inevitable and *natural* biological process to regarding it as a disease that itself should be treated. Reimagining ageing this way has the potential to significantly alter the way our society functions and conceptualizing ageing. For example, this reconceptualization can have the following rammifications:

- *Policy and social implications:* Disability research often delves into policies that impact people with disabilities. Considering ageing as a disease itself has the potential to impact healthcare policies, social support systems, and societal attitudes toward ageing and people who are elderly.
- *Consequences for quality of life and well-being:* Within disability studies, there is an emphasis on enhancing the quality of life and well-being of people with disabilities. Reimagining ageing as a disease could lead to interventions, research,

and societal changes aimed at improving the lives of ageing individuals, like efforts made for people with other disabilities.
- *Ramifications for healthcare:* By framing biological ageing as a disease that contributes to age-associated diseases, there might be a shift in healthcare priorities. More attention may be directed at preventive measures, early interventions, and treatments targeting the biological processes of ageing itself.

1.5 Intended Audience of This Book

Below are several audiences who would be interested in reading this book:

- *Researchers*: Researchers in the field of ageing and age-associated diseases, including those in academia, pharmaceutical companies, and biotech startups, would find this book to be of interest. It provides a comprehensive and contemporary overview of the latest research about senescent cells and senotherapies.
- *Clinicians*: Clinicians who treat patients with age-associated diseases would find this book useful. It provides an in-depth understanding of the impact of senescent cells in the development of age-associated diseases and research about possible drug candidates.
- *Students*: Students studying biology, medicine, or related fields would also be interested in this book. It provides a basic introduction to the topic of cellular senescence and senotherapy.
- *The public*: The public, especially those interested in maintaining their health and longevity, would find this book useful. It provides insights into the mechanisms of cellular senescence and age-associated diseases. It could be a valuable resource for anyone interested in improving their health and extending their lifespan.

Overall, anyone interested in the field of ageing research and therapeutic interventions that extend lifespan and healthspan, by preventing or delaying the onset of age-associated diseases, would benefit from reading this book.

1.6 Pedagogical Features of This Book

Different readers learn in different ways. For example, some derive an understanding of concepts after reading text, while others understand concepts after viewing an illustration. To cater for different readers, this book contains a combination of visual diagrams, quotes, and information in tabular form. For those who are interested in learning more about senotherapeutics, and other concepts in this book, a list of additional readings are presented at the end of Chaps. 1, 2, and 3. These pedagogical features ensure that the concepts embedded within this book can be understood by a wide audience.

1.7   Description of Upcoming Chapters

1.7.1   Chapter 2: The Accumulation of Senescent Cells and Disease

As biological ageing progresses, the acclamation of senescent cells exceeds an optimum therapeutic threshold, leading to an increased possibility of age-associated diseases and occurrence of disabilities. Chapter 2 presents the findings in the literature that explains that the cause of many age-associated diseases and disabilities is the accumulation of senescent cells beyond an optimum level. The diseases explored in this chapter include type 2 diabetes, type 1 diabetes, idiopathic pulmonary fibrosis, obesity-induced metabolic dysfunction, viral infections, ophthalmic diseases, multiple sclerosis, Parkinson's disease, and obesity-related diseases.

1.7.2   Chapter 3: Current Research About Senotherapeutics

It is beyond the scope of this book to review the literature about the therapeutic effectiveness of each senotherapeutic medication that is being tested. Thus, the therapeutic potential of the senolytic cocktail dasatinib and quercetin (D + Q) on different age-associated diseases will be examined. D + Q was selected because it appears to be a senolytic medication that has been extensively researched. The senomorphic potential of metformin and rapamycin will then be reviewed. A summary of the literature about human trials of senotherapy will then be presented, followed by an overview of research opportunities in the future regarding senotherapy.

1.7.3   Chapter 4: Final Comments

The final chapter summarizes the contents of this book and leaves the reader with the impression that more research should be conducted on the broader economic and societal consequences following a possible widespread adoption of senotherapy.

Further Reading

Davan-Wetton, C. S. A., Pessolano, E., Perretti, M., & Montero-Melendez, T. (2021). Senescence under appraisal: Hopes and challenges revisited. *Cellular and Molecular Life Sciences, 78*(7), 3333–3354. https://doi.org/10.1007/s00018-020-03746-x

de Magalhães, J. P. (2021). Longevity pharmacology comes of age. *Drug Discovery Today, 26*(7), 1559–1562. https://doi.org/10.1016/j.drudis.2021.02.015

Jurk, D., & Passos, J. F. (2022). Senolytic drugs: Beyond the promise and the hype. *Mechanisms of Ageing and Development, 202*, 111631. https://doi.org/10.1016/j.mad.2022.111631

Khalil, R., Diab-Assaf, M., & Lemaitre, J. M. (2023). Emerging therapeutic approaches to target the dark side of senescent cells: New hopes to treat aging as a disease and to delay age-related pathologies. *Cells, 12*(6), 915. https://doi.org/10.3390/cells12060915

Kitaeva, K. V., Solovyeva, V. V., Blatt, N. L., & Rizvanov, A. A. (2024). Eternal youth: A comprehensive exploration of gene, cellular, and pharmacological anti-aging strategies. *International Journal of Molecular Sciences, 25*(1), 643. https://doi.org/10.3390/ijms25010643

Lorenzo, E. C., Torrance, B. L., & Haynes, L. (2023). Impact of senolytic treatment on immunity, aging, and disease. *Frontiers in Aging, 4*, 1161799. https://doi.org/10.3389/fragi.2023.1161799

Martín-Vicente, P., López-Martínez, C., Rioseras, B., & Albaiceta, G. M. (2024). Activation of senescence in critically ill patients: Mechanisms, consequences and therapeutic opportunities. *Annals of Intensive Care, 14*(1), 2. https://doi.org/10.1186/s13613-023-01236-4

Popescu, I., Deelen, J., Illario, M., & Adams, J. (2023). Challenges in anti-aging medicine-trends in biomarker discovery and therapeutic interventions for a healthy lifespan. *Journal of Cellular and Molecular Medicine, 27*(18), 2643–2650. https://doi.org/10.1111/jcmm.17912

Raffaele, M., & Vinciguerra, M. (2022). The costs and benefits of senotherapeutics for human health. *The Lancet. Healthy Longevity, 3*(1), e67–e77. https://doi.org/10.1016/S2666-7568(21)00300-7

Smer-Barreto, V., Quintanilla, A., Elliott, R. J. R., Dawson, J. C., Sun, J., Campa, V. M., Lorente-Macías, Á., Unciti-Broceta, A., Carragher, N. O., Acosta, J. C., & Oyarzún, D. A. (2023). Discovery of senolytics using machine learning. *Nature Communications, 14*(1), 3445. https://doi.org/10.1038/s41467-023-39120-1

Sun, Y., Li, Q., & Kirkland, J. L. (2022). Targeting senescent cells for a healthier longevity: The roadmap for an era of global aging. *Life Medicine, 1*(2), 103–119. https://doi.org/10.1093/lifemedi/lnac030

References

Benhamú, B., Martín-Fontecha, M., Vázquez-Villa, H., López-Rodríguez, M. L., & Ortega-Gutiérrez, S. (2022). New trends in aging drug discovery. *Biomedicine, 10*(8), 2006. https://doi.org/10.3390/biomedicines10082006

Borghesan, M., Hoogaars, W. M. H., Varela-Eirin, M., Talma, N., & Demaria, M. (2020). A senescence-centric view of aging: Implications for longevity and disease. *Trends in Cell Biology, 30*(10), 777–791. https://doi.org/10.1016/j.tcb.2020.07.002

Bulterijs, S., Hull, R. S., Björk, V. C., & Roy, A. G. (2015). It is time to classify biological aging as a disease. *Frontiers in Genetics, 6*, 205. https://doi.org/10.3389/fgene.2015.00205

Chaib, S., Tchkonia, T., & Kirkland, J. L. (2022). Cellular senescence and senolytics: The path to the clinic. *Nature Medicine, 28*(8), 1556–1568. https://doi.org/10.1038/s41591-022-01923-y

Guo, J., Huang, X., Dou, L., Yan, M., Shen, T., Tang, W., & Li, J. (2022). Aging and aging-related diseases: From molecular mechanisms to interventions and treatments. *Signal Transduction and Targeted Therapy, 7*(1), 391. https://doi.org/10.1038/s41392-022-01251-0

López-Otín, C., Blasco, M. A., Partridge, L., Serrano, M., & Kroemer, G. (2013). The hallmarks of aging. *Cell, 153*(6), 1194–1217. https://doi.org/10.1016/j.cell.2013.05.039

Pignolo, R. J., Passos, J. F., Khosla, S., Tchkonia, T., & Kirkland, J. L. (2020). Reducing senescent cell burden in aging and disease. *Trends in Molecular Medicine, 26*(7), 630–638. https://doi.org/10.1016/j.molmed.2020.03.005

Schmauck-Medina, T., Molière, A., Lautrup, S., Zhang, J., Chlopicki, S., Madsen, H. B., Cao, S., Soendenbroe, C., Mansell, E., Vestergaard, M. B., Li, Z., Shiloh, Y., Opresko, P. L., Egly, J. M., Kirkwood, T., Verdin, E., Bohr, V. A., Cox, L. S., Stevnsner, T., Rasmussen, L. J., et al. (2022). New hallmarks of ageing: A 2022 Copenhagen ageing meeting summary. *Aging, 14*(16), 6829–6839. https://doi.org/10.18632/aging.204248

Schmitt, R. (2017). Senotherapy: Growing old and staying young? *Pflugers Archiv: European Journal of Physiology, 469*(9), 1051–1059. https://doi.org/10.1007/s00424-017-1972-4

Turrini, S., Wong, B., Eldaief, M., Press, D. Z., Sinclair, D. A., Koch, G., Avenanti, A., & Santarnecchi, E. (2023). The multifactorial nature of healthy brain ageing: Brain changes, functional decline and protective factors. *Ageing Research Reviews, 88*, 101939. https://doi.org/10.1016/j.arr.2023.101939

von Kobbe, C. (2019). Targeting senescent cells: Approaches, opportunities, challenges. *Aging, 11*(24), 12844–12861. https://doi.org/10.18632/aging.102557

Zhang, Q., Li, S., Chen, F., Zeng, R., & Tong, R. (2022). Targeted delivery strategy: A beneficial partner for emerging senotherapy. *Biomedicine & Pharmacotherapy, 155*, 113737. https://doi.org/10.1016/j.biopha.2022.113737

Chapter 2
The Accumulation of Senescent Cells and Diseases

Abstract This chapter explains the main findings in the literature about the impact of senescent cell accumulation on the development of age-associated diseases. The diseases explored in this chapter include type 2 diabetes, type 1 diabetes, idiopathic pulmonary fibrosis, obesity-induced metabolic dysfunction, viral infections, ophthalmic diseases, multiple sclerosis, Parkinson's disease, obesity-related diseases, non-alcoholic fatty liver disease, non-alcoholic steatohepatitis, and periodontitis. The purpose of this chapter is to present the findings in the literature that show that the accumulation of senescent cells can contribute to the development of age-associated diseases.

Keywords Age-associated diseases · Cellular senescence · Fatty liver disease · Idiopathic pulmonary fibrosis (IPF) · Multiple sclerosis · Obesity · Ophthalmic disease · Parkinson's disease · Periodontitis · Senescent cells · Type 2 diabetes · Viral infections

2.1 Causational Relationship Between Senescent Cell Accumulation and Age-Associated Diseases

2.1.1 Type 2 Diabetes

The accumulation of senescent cells can contribute to the development of type 2 diabetes by impacting the function of pancreatic beta cells. The metabolic changes that occur with diabetes, such as high blood sugar, can also stimulate the formation of senescent cells. This creates a feedback cycle in which senescent cells cause the development of diabetes and then diabetes causes the development of more senescent cells. Targeting the development of cellular senescence with senomorphics and reducing the accumulation of senescent cells with senolytics may be a more effective strategy at treating diabetes and its complications than current treatments that only address existing tissue damage. Therefore, senescent cells and the senescence-associated secretory phenotype (SASP) that they produce represent a promising

© The Author(s), under exclusive license to Springer Nature Singapore Pte Ltd. 2024

G. Bennett, *Senotherapy*, SpringerBriefs in Modern Perspectives on Disability Research, https://doi.org/10.1007/978-981-97-3637-9_2

area of research for the prevention and treatment of type 2 diabetes and its complications (Aguayo-Mazzucato et al., 2019; Narasimhan et al., 2021; Palmer et al., 2015; Palmer & Kirkland, 2016). However, there are two primary obstacles that currently hinder the clinical implementation of this technique: first, the underlying molecular mechanisms of cellular senescence in various organs are not yet fully comprehended, and second, it is necessary to determine the distinct impact of eliminating senescent cells in each individual organ (Iwasaki et al., 2023). As Palmer et al. (2015, p. 2295) explained:

> Senescent cell burden increases in aging and obesity and may play a role in causing or exacerbating type 2 diabetes. In turn, features of diabetes may cause an increase in senescent cell number, which would further promote chronic inflammation and initiate a vicious cycle of senescent cell formation in multiple tissues. This increased senescent cell burden may play a role in the tissue damage that contributes to diabetes complications. It will be crucial to explore the extent and characteristics of metabolic dysfunction in animal models with a high senescent cell burden in order to determine the effect of senescent cells on metabolic dysfunction, including insulin resistance. Shared mechanisms between the development of insulin resistance and cellular senescence indicate that perhaps by clearing senescent cells, both the metabolic components and complications of diabetes could be ameliorated. Experiments to test this possibility, such as in animal models that allow senescent cell clearance, will be extremely important in establishing whether senescent cells are a therapeutic target in diabetes. ... Glucose-lowering medications have limited effects on ameliorating existing tissue damage and might not be very effective in reducing senescent cell burden. Senolytics or SASP-protective agents, used alone or in conjunction with current glucose lowering therapies, may be a way to delay, prevent, alleviate, or even treat hitherto resistant complications of diabetes.

2.1.2 Type 1 Diabetes

Type 1 diabetes is an autoimmune disease that results in high blood sugar due to a loss of insulin-producing beta cells in the pancreas. Thompson et al.'s (2019) research showed that during the natural progression of type 1 diabetes, senescent beta cells emit SASPs. These cells are not removed by the immune system and the immune system instead contributes to the development of additional beta senescent cells. Eliminating these cells might stop the immune-mediated destruction of beta cells and prevent the continual onset of type 1 diabetes. If this proposition is true, then beta cell senescence plays a significant role in the development of type 1 diabetes and that targeting senescent beta cells could be a potential new therapy for type 1 diabetes (Thompson et al., 2019). As Thompson et al. (2019, p, 1057) explained:

> Although current therapies for T1D [type 1 diabetes] have focused almost exclusively on disabling the immune cell attack on beta cells, our studies suggest that defects in the immune surveillance and clearance of SASP beta cells plays a crucial and previously unrecognized role in the pathogenesis of the disease. The clearance of SASP beta cells with senolytic agents could be viewed as compensating for the failure of the immune response to SASP cell surveillance in T1D. ... Targeting SASP beta cells using senolytic

drugs to halt the disease and preserve functional beta cell mass provides a new paradigm for T1D therapy.

2.1.3 Idiopathic Pulmonary Fibrosis (IPF)

Idiopathic pulmonary fibrosis (IPF) is a progressive and fatal lung disorder of unknown cause and with limited treatment options. The incidence and severity of IPF increases with age, suggesting that the ageing process itself plays a role in its development. Senescent cell accumulation has been linked to the biological process of ageing and several age-associated diseases, including IPF. Senescent cells have been found in IPF lungs and in experimental models of lung fibrosis. Human experiments have shown that using senolytic medications to reduce senescent cell accumulation can improve lung function, which can possibly reverse the progression of IPF (Justice et al., 2019). These findings suggest that the accumulation of senescent cells can have an impact on the development of IPF and that targeting senescent cells may be a novel approach to treating this age-associated disease (Han et al., 2023; Liu & Liu, 2020). As Liu and Liu (2020, p. 5) concluded:

> *Cellular senescence has emerged as an important mechanism underlying the pathogenesis of IPF* [Idiopathic pulmonary fibrosis]. *More studies are needed to improve our understanding of the molecular mechanisms leading to cellular senescence in IPF lungs and the mechanisms whereby senescent cells contribute to the development of lung fibrosis. Most importantly, the efficacy of senotherapy for the treatment of IPF warrants further investigation.*

2.1.4 Viral Infections

Evidence suggests that viral-induced cellular senescence combined with preexisting cellular senescence burden in the elderly can worsen the severity of viral infections, trigger excessive age-associated inflammation, and lead to multi-organ damage or dysfunction, which ultimately results in higher mortality. Consequently, there is research about the positive effects of senescence-targeted drugs in treating viral infectious diseases among the elderly. For example, Li et al.'s (2023) review explored the relationship between cellular senescence and viral infection, as well as the significance of senotherapy in treating such infections. They concluded that:

> *Senescence and viral infections interact in a reciprocal relationship. In general, viral infections can induce senescence and increase the susceptibility and severity of viral infections via multiple mechanisms, such as immunodeficiency, mitochondrial dysfunction, SASP secretion, pre-activated macrophages, over-recruitment of immune cells, and accumulation of innate immune cells with trained immunity. In the elderly, virus-induced senescence, in addition to their pre-existing senescent condition, is believed to aggravate the underlying disease outcomes, but could be counteracted by senotherapeutics, which was shown to mitigate the severity of viral infections. (Li et al., 2023, p. 12)*

2.1.5   Ophthalmic Diseases

The role cellular senescence plays in the development of various ocular diseases has been explained. Soleimani et al. (2023) summarized our knowledge about this topic by conducting a literature search of PubMed using the search query "senescence OR aging AND eye disease OR ocular disease OR ophthalmic disease OR cornea OR glaucoma OR cataract OR retina." A total of 51 articles about cellular senescence and ocular diseases were identified and summarized. They concluded that the accumulation of senescence cells has been linked to multiple corneal and retinal pathologies, as well as glaucoma and cataracts. Senotheraputic medications that selectively target senescent cells in eyes can be used as a therapeutic approach. However, despite expanding rapidly, the literature on cellular senescence and ocular disease is relatively new. It is still debated whether cellular senescence contributes significantly to ocular diseases. To date, no human studies have demonstrated the benefits of senolytic or senomorphic therapies for restoring ocular functioning (Soleimani et al., 2023). As Soleimani et al. (2023, pp. 3077–3078) concluded:

> Senescence has been shown to underlie the pathogenesis of numerous ocular diseases including those of the ocular surface, lens, retina, and optic nerve. The overall literature on senescence and ocular disease is growing rapidly—over 100 papers are published annually on the topic since 2016. There is no shortage of studies which report correlations between senescence and the progression of disease states. A significant number of these demonstrate causative effects of senescence in promoting disease. Fewer still have tested the efficacy of senolytic therapies in treating or preventing ocular diseases in in vivo models. Finally, no studies exist to date which have demonstrated the benefits of senolytic therapies in human studies. There is therefore a critical unmet need to perform studies which examine the effects of senolytic therapies in ex vivo models, in vivo models, and eventually clinical trials for ocular pathologies. Future work which investigates the translational efficacy of senolytic agents would fulfill this niche and greatly accelerate the development of targeted therapies for ocular diseases.

2.1.6   Multiple Sclerosis

Ageing is a significant risk factor in the development of several neurodegenerative diseases, including multiple sclerosis. One of the underlying biological processes associated with ageing is the accumulation of senescent cells. Several stressors linked to the development of biological ageing and multiple sclerosis pathology can trigger cellular senescence accumulation, such as oxidative stress, mitochondrial dysfunction, cytokines, and replicative exhaustion. In multiple sclerosis, cellular senescence accumulation may contribute to disease progression. Papadopoulos et al. (2020) discussed the evidence linking cellular senescence to the development of multiple sclerosis and the potential role of senolytic and senomorphic medications in treating the progression of multiple sclerosis and providing neuroprotection. They concluded that:

A primary causative role of CS [Cellular senescent] in MS [Multiple sclerosis] is highly unlikely given the great diversity which characterizes aging-related neurodegenerative pathologies that have CS as a common feature. However, CS may be a shared mechanism, which substantially contributes to the pathogenesis and impact of neurodegenerative diseases and thereby may determine disease susceptibility, age at disease presentation and rate of progression. … Nevertheless, current evidence for a role of CS in disability progression in MS is intriguing but limited and indirect. Shedding light on CS and its role in neurodegeneration is essential to safely exploit it therapeutically. To facilitate these efforts a thorough histopathological investigation of post-mortem MS tissue at various disease stages and levels of disability would inform us of the extent, timing, particular cell types converted to senescence and all features of pathology associated with the accumulation of senescent cells. A more concise understanding of the biology of CS of neural cells, its triggers and mediators is required. (Papadopoulos et al., 2020, pp. 6–7)

In Oost et al.'s (2018) literature review, the possible link between cellular senescence and the progression of multiple sclerosis was discussed, and the potential use of senolytics as a treatment for multiple sclerosis was explored. Currently, there is no cure for multiple sclerosis and there are limited treatment options that can slow disease progression. Various cell types present in the central nervous system can become senescent and thus potentially contribute to the progression of multiple sclerosis. Based on the literature that they examined, Oost et al. proposed that cellular senescence can directly impact the progression of multiple sclerosis and that the administration of senolytics should be tested as a potential treatment approach.

2.1.7 Parkinson's Disease

Currently, over 10 million people worldwide live with Parkinson's disease. The accumulation of senescent cells in the brain may contribute to the development and progression of Parkinson's disease. Studies have shown that senescent cells can increase oxidative stress and neuroinflammation, both of which are implicated in Parkinson's disease. Kakoty et al. (2023) examined the literature about the impact of senescence cell accumulation on the development of Parkinson's disease, highlighting recent research advancements in the field of senolytics and their potential as a future pharmaceutical treatment for Parkinson's disease. They concluded that:

Owing to the prevalence rate of PD [Parkinson's disease] in elderly patients, it has become necessary for discovery of safe and effective anti-PD therapeutics. Bioactive compounds of natural origin have been found to be potent as well as considerably safe compared with synthetic drugs. In this context, senolytics hold a great potential to treat PD. The present review has described the mechanism through which the senolytics inhibit the progression of PD by killing SC. The preclinical evidence has shown good senolytic activity of Dasa, Quer, Fisetin and navitoclax. Further, some of the case studies have also been discussed in detail wherein the scientists have evaluated that successful killing of SC in animal models has shown amelioration in the pathogenesis of PD. Although the research is in its nascent stage, the promising results of existing studies have indicated that senolytics have a great potential ahead as anti-PD agents. (Kakoty et al., 2023, p. 7)

2.1.8   Obesity-Related Diseases

Reducing senescent cell burden may represent an effective strategy for treating obesity-related health problems and diseases. Senotherapeutic drugs may have the potential to be such a strategy to achieve this outcome. Narasimhan et al. (2022) examined the literature about the relationship between obesity and cellular senescence in various organs and explored the role of senescent cells in obesity-related health complications. They also reviewed the literature about the potential use of senotherapeutic drugs to target senescent cells as a treatment for these complications. They summarized that:

> The accumulation of SnCs [senescent cells] with age and disease accelerates aging. Obesity is a key driver of SnC accumulation, and the complications associated with obesity can be controlled by reducing the SnC burden. Thus, senotherapeutic drugs have the potential to be an effective therapeutic option. (Narasimhan et al., 2022, p. 537)

2.1.9   Non-alcoholic Fatty Liver Disease (NAFLD)

Engelmann and Tacke (2022) reviewed the evidence about the causational relationship between cellular senescence and the development of non-alcoholic fatty liver disease (NAFLD). They concluded that senescent cells may be a key factor in the development of NAFLD and may contribute to intracellular fat accumulation, fibrosis, and inflammation due to the production of SASPs. A deeper understanding of the interplay between NAFLD and cellular senescence is necessary to identify new therapeutic targets for halting disease progression.

2.1.10   Non-alcoholic Steatohepatitis (NASH)

Bonnet et al. (2022) investigated if cellular senescence contributed to the creation of non-alcoholic steatohepatitis (NASH). They treated two human hepatocyte cell lines with drugs that induce senescence and found that these drugs led to the activation of senescence-associated markers and the development of hepatic steatosis. Despite inducing senescence, insulin signaling was increased in the cells, indicating that cellular senescence could play a causal role in the development of NASH pathogenesis by altering glucose and lipid metabolism. The results of this study suggest that targeting senescent cells could be a viable therapeutic strategy for the treatment of NASH (Bonnet et al., 2022). Bonnet et al. (2022, p. 13) concluded that:

> Taken together, our results show that cellular senescence in hepatocytes causes important changes in glucose and lipid metabolism, leading to increased steatosis, and increased glucose and TG production. Senescence may thus play a causal role in the development and progression of NAFLD, and targeting cellular senescence may represent a promising therapeutic approach for the treatment of NASH.

2.1.11   Periodontitis

Periodontitis is a disease characterized by chronic inflammation of the gums, which becomes more prevalent and severe in older individuals. Biological ageing is a major risk factor for the development of periodontitis, resulting in alveolar bone loss and tooth loss in elderly individuals. However, the mechanisms by which ageing contributes to periodontitis are not well understood. Recent research suggests that targeting the accumulation of cellular senescence may slow the ageing process and potentially alleviate age-associated diseases, such as periodontitis. Senescent cells accumulate in the alveolar bone and secrete SASPs, which interact with bacteria and exacerbate the chronic inflammation in the periodontal tissue, leading to increased alveolar bone loss. Chen et al. (2022) proposed that the progression of periodontitis can be slowed or reversed if the increase of senescent cells and the SASPs that they can emit can be thwarted.

2.2   Conclusion

The purpose of this chapter was to present the literature that explains the causational relationship between senescent cell accumulation and the development of age-associated diseases. The diseases explored in this chapter included type 2 diabetes, idiopathic pulmonary fibrosis, obesity-induced metabolic dysfunction, viral infections, and Parkinson's disease. An appraisal of the research examined suggests that the accumulation of senescent cells beyond a safe therapeutic threshold contributes to the development of age-associated diseases and disabilities. Consequently, reducing this accumulation to a safe level is a target for therapeutic intervention, and this therapeutic approach will be the focus of the next chapter.

Further Reading

Borghesan, M., Hoogaars, W. M. H., Varela-Eirin, M., Talma, N., & Demaria, M. (2020). A senescence-centric view of aging: Implications for longevity and disease. *Trends in Cell Biology, 30*(10), 777–791. https://doi.org/10.1016/j.tcb.2020.07.002

Khalil, R., Diab-Assaf, M., & Lemaitre, J. M. (2023). Emerging therapeutic approaches to target the dark side of senescent cells: New hopes to treat aging as a disease and to delay age-related pathologies. *Cells, 12*(6), 915. https://doi.org/10.3390/cells12060915

References

Aguayo-Mazzucato, C., Andle, J., Lee, T. B., Jr., Midha, A., Talemal, L., Chipashvili, V., Hollister-Lock, J., van Deursen, J., Weir, G., & Bonner-Weir, S. (2019). Acceleration of β cell aging determines diabetes and senolysis improves disease outcomes. *Cell Metabolism, 30*(1), 129–142.e4. https://doi.org/10.1016/j.cmet.2019.05.006

Bonnet, L., Alexandersson, I., Baboota, R. K., Kroon, T., Oscarsson, J., Smith, U., & Boucher, J. (2022). Cellular senescence in hepatocytes contributes to metabolic disturbances in NASH. *Frontiers in Endocrinology, 13*, 957616. https://doi.org/10.3389/fendo.2022.957616

Chen, S., Zhou, D., Liu, O., Chen, H., Wang, Y., & Zhou, Y. (2022). Cellular senescence and periodontitis: Mechanisms and therapeutics. *Biology, 11*(10), 1419. https://doi.org/10.3390/biology11101419

Engelmann, C., & Tacke, F. (2022). The potential role of cellular senescence in non-alcoholic fatty liver disease. *International Journal of Molecular Sciences, 23*(2), 652. https://doi.org/10.3390/ijms23020652

Han, S., Lu, Q., & Liu, X. (2023). Advances in cellular senescence in idiopathic pulmonary fibrosis (Review). *Experimental and Therapeutic Medicine, 25*(4), 145. https://doi.org/10.3892/etm.2023.11844

Iwasaki, K., Abarca, C., & Aguayo-Mazzucato, C. (2023). Regulation of cellular senescence in type 2 diabetes mellitus: From mechanisms to clinical applications. *Diabetes & Metabolism Journal, 47*(4), 441–453. https://doi.org/10.4093/dmj.2022.0416

Justice, J. N., Nambiar, A. M., Tchkonia, T., LeBrasseur, N. K., Pascual, R., Hashmi, S. K., Prata, L., Masternak, M. M., Kritchevsky, S. B., Musi, N., & Kirkland, J. L. (2019). Senolytics in idiopathic pulmonary fibrosis: Results from a first-in-human, open-label, pilot study. *eBioMedicine, 40*, 554–563. https://doi.org/10.1016/j.ebiom.2018.12.052

Kakoty, V., Kalarikkal Chandran, S., Gulati, M., Goh, B. H., Dua, K., & Kumar Singh, S. (2023). Senolytics: Opening avenues in drug discovery to find novel therapeutics for Parkinson's disease. *Drug Discovery Today, 28*(6), 103582. https://doi.org/10.1016/j.drudis.2023.103582

Li, Z., Tian, M., Wang, G., Cui, X., Ma, J., Liu, S., Shen, B., Liu, F., Wu, K., Xiao, X., & Zhu, C. (2023). Senotherapeutics: An emerging approach to the treatment of viral infectious diseases in the elderly. *Frontiers in Cellular and Infection Microbiology, 13*, 1098712. https://doi.org/10.3389/fcimb.2023.1098712

Liu, R. M., & Liu, G. (2020). Cell senescence and fibrotic lung diseases. *Experimental Gerontology, 132*, 110836. https://doi.org/10.1016/j.exger.2020.110836

Narasimhan, A., Flores, R. R., Robbins, P. D., & Niedernhofer, L. J. (2021). Role of cellular senescence in type II diabetes. *Endocrinology, 162*(10), bqab136. https://doi.org/10.1210/endocr/bqab136

Narasimhan, A., Flores, R. R., Camell, C. D., Bernlohr, D. A., Robbins, P. D., & Niedernhofer, L. J. (2022). Cellular senescence in obesity and associated complications: A new therapeutic target. *Current Diabetes Reports, 22*(11), 537–548. https://doi.org/10.1007/s11892-022-01493-w

Oost, W., Talma, N., Meilof, J. F., & Laman, J. D. (2018). Targeting senescence to delay progression of multiple sclerosis. *Journal of Molecular Medicine (Berlin, Germany), 96*(11), 1153–1166. https://doi.org/10.1007/s00109-018-1686-x

Palmer, A. K., & Kirkland, J. L. (2016). Aging and adipose tissue: Potential interventions for diabetes and regenerative medicine. *Experimental Gerontology, 86*, 97–105. https://doi.org/10.1016/j.exger.2016.02.013

Palmer, A. K., Tchkonia, T., LeBrasseur, N. K., Chini, E. N., Xu, M., & Kirkland, J. L. (2015). Cellular senescence in type 2 diabetes: A therapeutic opportunity. *Diabetes, 64*(7), 2289–2298. https://doi.org/10.2337/db14-1820

Papadopoulos, D., Magliozzi, R., Mitsikostas, D. D., Gorgoulis, V. G., & Nicholas, R. S. (2020). Aging, cellular senescence, and progressive multiple sclerosis. *Frontiers in Cellular Neuroscience, 14*, 178. https://doi.org/10.3389/fncel.2020.00178

Soleimani, M., Cheraqpour, K., Koganti, R., & Djalilian, A. R. (2023). Cellular senescence and ophthalmic diseases: Narrative review. *Graefe's Archive for Clinical and Experimental Ophthalmology, 261*(11), 3067–3082. https://doi.org/10.1007/s00417-023-06070-9

Thompson, P. J., Shah, A., Ntranos, V., Van Gool, F., Atkinson, M., & Bhushan, A. (2019). Targeted elimination of senescent beta cells prevents type 1 diabetes. *Cell Metabolism, 29*(5), 1045–1060.e10. https://doi.org/10.1016/j.cmet.2019.01.021

Chapter 3
Current Research About Senotherapeutics

Abstract In the previous chapter, it was proposed that when the quantity of senescent cells exceeds a safe therapeutic threshold, the prospect of age-associated diseases increases. The goal of senolytics is to destroy senescent cells so that their volume remains below this threshold. However, below this threshold, senescent cells provide safe transitory levels of senescence-associated secretory phenotypes (SASPs), which assist with wound healing and tumor suppression. Senomorphics, on the other hand, can prevent senescent cells from making other cells senescent via constant exposure to SASPs. A review of the literature about senotherapeutics revealed a diverse range of senolytic and senomorphic candidates. It is beyond the scope of this book to examine the literature about the merits of each candidate. Thus, the literature about the senolytic potential of dasatinib and quercetin (D + Q) and the senomorphic potential of metformin and rapamycin are reviewed. Despite the potential of senotherapy to extend both human lifespan and healthspan, there remains many opportunities to explore this emerging field of medicine. To emphasize this point, this chapter concludes with a series of research topics and questions about senotherapeutics that remain to be investigated in the future.

Keywords Cancer · Cardiovascular disease · Chronic kidney disease (CKD) · Dasatinib and quercetin (D + Q) · Future research opportunities · Human trials · Metabolic disorders · Metformin · Non-alcoholic fatty liver disease (NAFLD) · Rapamycin · Senotherapy

3.1 Senolytic Potential of Dasatinib and Quercetin (D + Q)

3.1.1 Learning and Memory

Typically, as a person's biological age increases, their cognitive abilities decline. Krzystyniak et al. (2022) tested if the senolytic cocktail dasatinib and quercetin (D + Q) could reverse this decline. They found that after 8 weeks of D + Q

G. Bennett, *Senotherapy*, SpringerBriefs in Modern Perspectives on Disability Research, https://doi.org/10.1007/978-981-97-3637-9_3

treatment, male Wistar rats demonstrated improved learning and memory abilities. They also noted that these beneficial effects on cognitive function were long-lasting and persisted even after treatment was discontinued (Krzystyniak et al., 2022).

3.1.2 Reduced Adipose Tissue Inflammation and Improved Systemic Metabolic Function

Research indicates that removing senescent cells can alleviate chronic inflammation and its associated dysfunction and diseases. However, the impact of this intervention on metabolic function in old age is not well understood. Islam et al. (2023) reported that they did not observe a robust effect of D + Q treatment on senescence and inflammatory senescence-associated secretory phenotype (SASP) markers in the liver and skeletal muscle of old mice. However, they did find that D + Q treatment improved fasting blood glucose and glucose tolerance and lowered hepatic gluconeogenesis. These findings suggest that for aged mice D + Q treatment can mitigate adipose tissue inflammation and improve systemic metabolic function. Based on these results, the development of therapeutic agents to combat metabolic dysfunction and diseases in old age seems plausible (Islam et al., 2023).

3.1.3 Muscle Wasting in Chronic Kidney Disease (CKD)

Huang et al. (2023) examined if SASPs emitted by senescent cells influenced the development of muscle wasting and weakness associated with chronic kidney disease (CKD). In their study, they induced CKD in mice using a method called five-sixths nephrectomy. They then measured kidney function and muscle size and function and evaluated markers of atrophy, inflammation, and senescence using various techniques such as immunohistochemistry, immunoblots, or qPCR. To investigate the role of senescence, the researchers administered D + Q orally to mice for 8 weeks. They found that the administration of D + Q over 8 weeks eliminated the elevated markers of senescence and reduced high levels of SASPs. In addition, D + Q treatment improved skeletal muscle weight and grip function in mice with CKD. These findings suggest that senescent cells play a role in CKD-induced muscle atrophy and weakness and that D + Q may be a promising therapeutic approach for CKD-associated muscle wasting. Furthermore, these findings may provide new avenues for therapeutic intervention to improve muscle health and function in individuals with CKD (Huang et al., 2023).

3.1.4 Age-Dependent Intervertebral Disc Degeneration

It has been proposed that the accumulation of senescent cells can contribute to the development of disc degeneration. To explore this causational link, Novais et al. (2021) investigated if a combination of D + Q removed the accumulation of senescent cells and in doing so reversed the progression of disc degeneration among a sample of mice. For their sample, they included mice who were of different ages (i.e., 6, 14, and 18 months). The data that they collected showed that the 6- and 14-month cohorts who were treated with D + Q had lower incidences of disc degeneration and contributed to a significant decrease in senescence markers (i.e., p16INK4a, p19ARF) and molecules associated with inflammation (i.e., IL-6 and MMP13). Novais et al.'s findings suggest that senolytics, such as D + Q, may be an effective strategy for mitigating age-associated intervertebral disc degeneration (Novais et al., 2021). As they suggested:

> *In summary, systemic administration of D+Q posits an exciting therapeutic approach to treating disc degeneration, without the inherent risks associated with invasive surgical interventions. The potential benefits of D+Q treatment include alleviation of disc degeneration, reduction in systemic inflammation, and improved physical condition during aging.*
> (Novais et al., 2021, p. 15)

3.1.5 Non-alcoholic Fatty Liver Disease (NAFLD)

Non-alcoholic fatty liver disease (NAFLD) is a disease that can progress to hepatocellular carcinoma (HCC) (see Fig. 3.1) (Engelmann & Tacke, 2022).

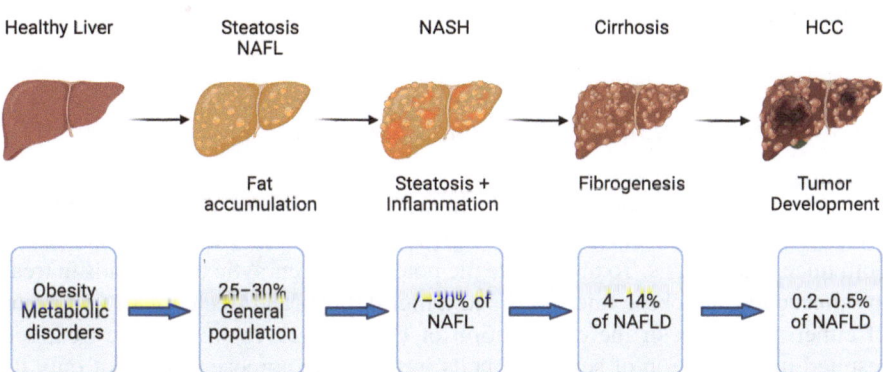

Fig. 3.1 Disease progression in fatty liver disease. Notes: Obesity and metabolic disorders represent risk factors for NAFLD, which starts with fat accumulation in hepatocytes. Further disease progression is characterized by inflammation (NASH) and subsequent fibrogenesis leading to cirrhosis. NASH with and without cirrhosis bears the risk of developing hepatocellular carcinoma (HCC). (Source: Engelmann & Tacke, 2022, p. 2)

As the concentration of senescent cells increases, the probability of tissue degeneration and HCC increases. Previous studies have shown that D + Q treatment can reduce NAFLD in mice. Raffaele et al. (2021) fed mice a diet that promoted liver inflammation and tumor development over a period of 9 months. They then divided the mice into four experimental groups: (1) a control group, (2) a group treated with D + Q, (3) a group undergoing the low dose of diethylnitrosamine (an extremely potent liver carcinogen for rodents) (DEN) and a high-fat diet (HFD) regiment, and (4) a group undergoing the DEN/HFD regimen with D + Q treatment. At the end of the 9-month regimen, the researchers used histopathology, qPCR, and immunoblotting techniques to assess the extent of liver damage and cell senescence. It was discovered that D + Q treatment had a negative effect on the progression of liver disease in the DEN/HFD mouse group, leading to a slight increase in both histological damage and tumorigenesis. D + Q also did not decrease the volume of senescent cells (Raffaele et al., 2021).

3.1.6   Cardiovascular Diseases

It is unknown if senolytic drugs can promote heart repair and regeneration after a heart attack. Salerno et al. (2022) gave D + Q treatment to six female mice aged 22–24 months that had a heart attack. D + Q improved overall heart function and performance and reduced the accumulation of senescent cells in the heart. While this study concluded that removing senescent cells is an important therapeutic target for repairing the hearts of female mice after a heart attack, translational research for human subjects remains to be conducted (Salerno et al., 2022).

3.1.7   Obesity-Induced Metabolic Disorders

The accumulation of senescent cells in white adipose tissue is associated with inflammation and the development of glucose intolerance and type 2 diabetes in both mice and humans. To investigate the potential of senolytic compounds in treating this condition, obese mice were fed a high-fat diet and treated with five cycles of either navitoclax or the combination of D + Q over 16 weeks. The treatment resulted in a reduction of senescent cells in the white adipose tissue, but only the D + Q treatment showed a short-term improvement in insulin sensitivity and glucose tolerance during cycles 3 and 4. However, this effect disappeared by the fifth cycle. These results suggest that both navitoclax and D + Q can temporarily improve metabolic disorders induced by obesity (Sierra-Ramirez et al., 2020). The therapeutic potential of D + Q or navitoclax for improving metabolic homeostasis among obese humans is currently unexamined.

3.2   Senomorphic Potential of Metformin

Metformin, a medication that is primarily used to assist in the regulation of blood glucose levels for people with diabetes, was first extracted from the French lilac plant during the 1600s (Triggle et al., 2022) (see Appendix 1). Since then, the therapeutic benefits that metformin provides, by altering the hallmarks of ageing, have been the subject of research (Chen et al., 2022) (see Fig. 3.2) (see Appendix 2).

Literature reviews by Campbell et al. (2017) and Mohammed et al. (2021) have shown that metformin has the potential to extend human lifespan and healthspan. Campbell et al.'s systematic review aimed to determine if metformin might have a protective effect against ageing in humans. They searched PubMed and Embase, as well as databases of unpublished studies, to identify relevant research that investigated the effect of metformin on all-cause mortality or age-associated diseases in nondiabetic and diabetic populations. Out of 260 full texts that were reviewed, 53 studies met their inclusion criteria. Results from the eligible research suggested that those who consume metformin had lower all-cause mortality than nondiabetics and reduced mortality compared to diabetics receiving non-metformin therapies, such as insulin. Additionally, those who took metformin had a lower incidence of cancer compared to nondiabetics and reduced cardiovascular disease compared to diabetics receiving non-metformin therapies or insulin. They concluded that:

> The findings reported in this systematic review remain preliminary generalisations, primarily making use of existing observational evidence collected for other purposes to investigate the credibility of the hypothesis that the insulin sensitiser metformin may extend the health and lifespans of people from the non-diabetic population. Differences in baseline characteristics were found which had the potential to bias results both towards positive findings (where metformin users were compared to other diabetics) as well as away from (where metformin users were compared to non-diabetics). While they should not be overstated, the apparent association with reductions in all-cause mortality and diseases of ageing found through meta-analysis do support this hypothesis, and metformin should be investigated as an intervention for ageing in future clinical trials. (Campbell et al., 2017, p. 42)

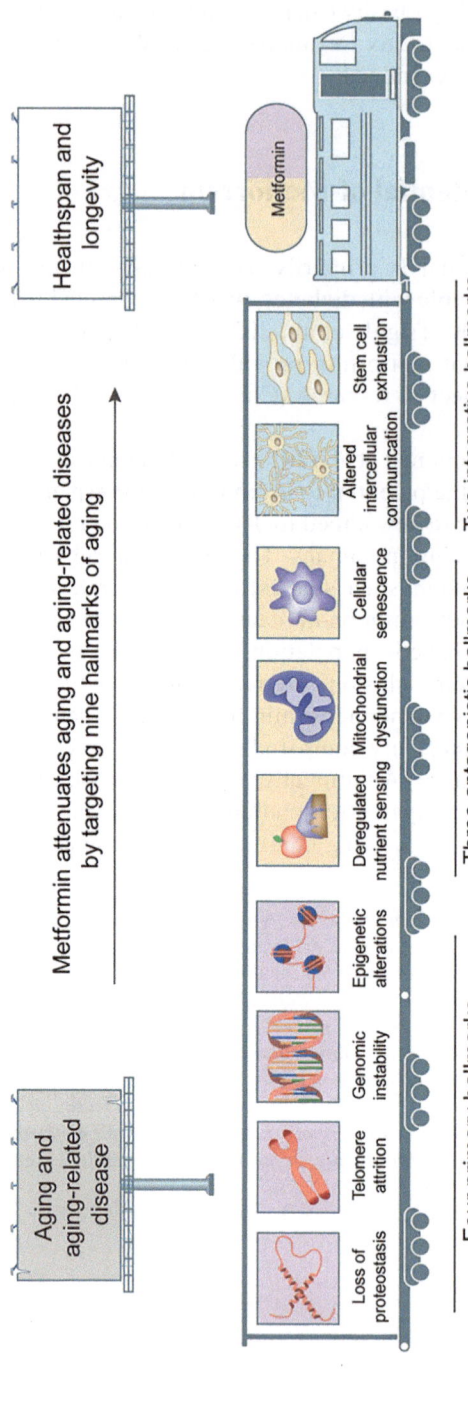

Fig. 3.2 Targets of metformin among the hallmarks of ageing. Notes: Metformin attenuates ageing and ageing-related diseases by targeting nine hallmarks of ageing, including (1) four primary hallmarks (loss of proteostasis, telomere attrition, genomic instability, and epigenetic alterations), (2) three antagonistic hallmarks (deregulated nutrient sensing, mitochondrial dysfunction, and cellular senescence), and (3) two integrative hallmarks (altered intercellular communication and stem cell exhaustion). (Source: Chen et al. 2022, p. 2730)

In a more contemporary study, Mohammed et al. evaluated the quality and results of 18 studies that examined the effects of metformin on lifespan and healthspan in humans or other species. They discovered that some studies concluded that metformin can be a geroprotector and provide anti-ageing benefits, such as reducing the possibility of neurological deterioration. However, metformin's potential to increase lifespan remains inconclusive and contentious. As Mohammed et al. (2021, p. 17) concluded:

> In this review we have also highlighted and critiqued some of the key clinical and laboratory-based studies that provide data supportive of the hypothesis that metformin, independent of its anti-hyperglycemic actions, has benefits that in principle can slow cellular aging and enhance healthspan and lifespan. Metformin, via its direct protective effects on vascular function, may slow the aging process via improved blood flow and provide protection against age-related cognitive decline. However, not all of the data is supportive and metformin, as shown in C. elegans and mice, may be less effective, or ineffective, in older humans. We have also stressed that, based on the pharmacokinetic properties of metformin, caution is needed before extrapolating from in vitro cell-based studies done with comparatively high metformin concentrations to clinical effectiveness with plasma concentrations in the range of 20 micromolar or lower.

Mohammed et al. and Campbell et al. suggest that metformin may have a *gero-protective effect* by potentially altering the mechanisms that impact the hallmarks of ageing, which consequently extends both lifespan and healthspan. Despite its potential to slow the biological processes that cause ageing, as illustrated by the hallmarks of ageing resembling those who are biological young, more research about metformin is needed. To address this research deficit, two large clinical studies are being conducted, *Targeting Aging with Metformin* (TAME) and *Metformin in Longevity Study* (MILES) (Triggle et al., 2022). Additionally, as shown in https://www.clinicaltrials.gov, the US government is funding several studies into the senotheraputic/senomorphic potential of metformin (see Appendix 3).

3.3   Senomorphic Potential of Rapamycin

Rapamycin, and its derivatives (i.e., rapalogs), can inhibit mTOR, a mechanism of the ageing process. With the intention of summarizing the findings in the literature about the effects of rapamycin and its derivatives on mitigating the severity of age-associated diseases, Lee et al. (2024) identified 19 studies from 18,400 studies retrieved from five databases (see Appendix 4). They did not identify any studies about the therapeutic benefits of rapamycin and its derivatives on respiratory, digestive, renal, and reproductive systems. However, rapamycin and its derivatives were found to improve cardiovascular, immune, and integumentary systems among healthy people and people with age-associated diseases. Additionally, endocrine, muscular, and neurological systems did not improve after consuming rapamycin or its derivatives (see Figs. 3.3 and 3.4).

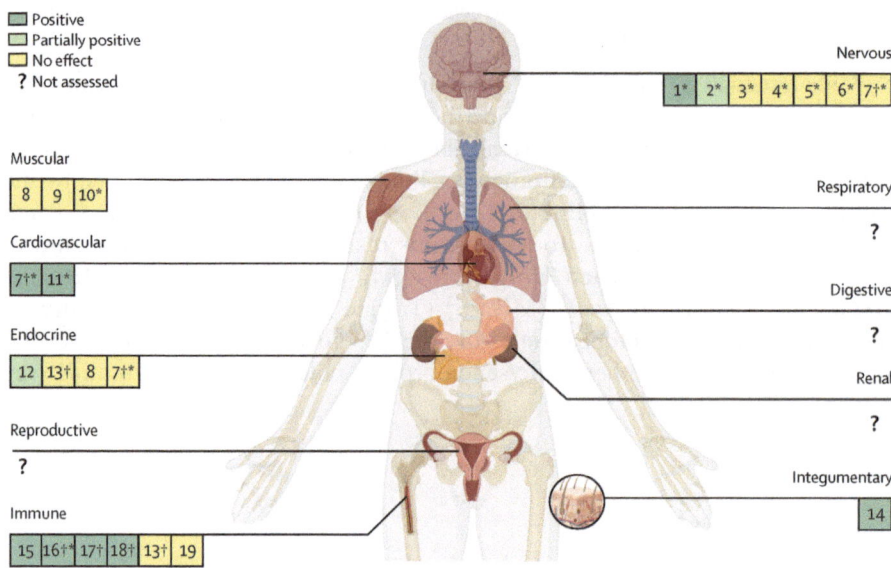

Fig. 3.3 Summary of the overall effect of rapamycin and its derivatives on physiological systems in healthy individuals and individuals with ageing-related diseases. Notes: *Studies on individuals with age-related diseases. †Studies using rapamycin derivatives. (1) Minturn et al., 2021, (2) Petrou et al., 2014, (3) Nussenblatt et al., 2010, (4) Palma et al., 2022, (5) Gensler et al., 2018, (6) Dugel et al., 2012, (7) Seyfarth et al., 2013, (8) Drummond et al., 2009, (9) Dickinson et al., 2013, (10) Gundermann et al., 2014, (11) Boni et al., 2012, (12) Krebs et al., 2007, (13) Hörbelt et al., 2020, (14) Chung et al., 2019, (15) Wen et al., 2019, (16) Bruyn et al., 2008, (17) Mannick et al., 2021, (18) Mannick et al., 2014, (19) Kraig et al., 2018. (Source: Lee et al., 2024, p. e156)

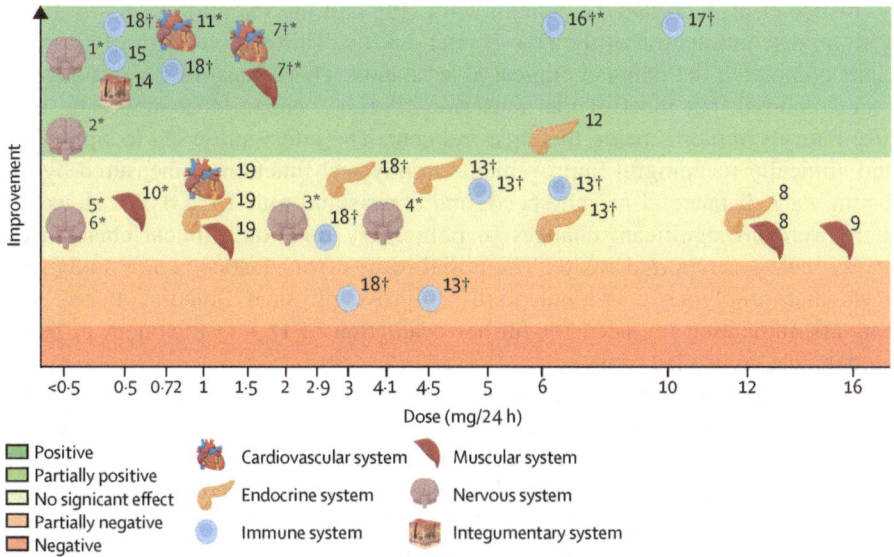

Fig. 3.4 Rapamycin and its derivatives; dose-effect relationship on physiological systems in healthy individuals and individuals with ageing-related diseases. Notes: *Studies on individuals with age-related diseases. †Studies using rapamycin derivatives. (1) Minturn et al., 2021, (2) Petrou et al., 2014, (3) Nussenblatt et al., 2010, (4) Palma et al., 2022, (5) Gensler et al., 2018, (6) Dugel et al., 2012, (7) Seyfarth et al., 2013, (8) Drummond et al., 2009, (9) Dickinson et al., 2013, (10) Gundermann et al., 2014, (11) Boni et al., 2012, (12) Krebs et al., 2007, (13) Hörbelt et al., 2020, (14) Chung et al., 2019, (15) Wen et al., 2019, (16) Bruyn et al., 2008, (17) Mannick et al., 2021, (18) Mannick et al., 2014, (19) Kraig et al., 2018. (Source: Lee et al., 2024, p. e156)

3.4   Human Trials of Senotherapeutic Candidates

Currently, there are no senolytic and senomorphic medications that have been approved for medical use in humans. However, there are a few human senolytic trials occurring (Hickson et al., 2019; Justice et al., 2019; Nambiar et al., 2023).

The accumulation of senescent cells contributes to the development of age-associated diseases. Reducing this accumulation has not been a focus for therapeutic intervention. Idiopathic pulmonary fibrosis (IPF) is an example of a fatal cellular senescence-associated disease that progresses over time. Studies on mice have shown that selectively removing senescent cells using a combination of D + Q can alleviate IPF-related dysfunction. Justice et al. (2019) evaluated the feasibility of a senolytic intervention for human participants with IPF ($n = 14$). Their open-label study involved intermittent administration of D + Q (dasatinib, 100 mg/day; quercetin, 1250 mg/day) for 3 days per week over a period of 3 weeks. The primary focus of this study was the retention rates and completion rates of the planned clinical assessments. The secondary focus was the safety of the intervention and the changes in functional and reported health measures (Justice et al., 2019).

Justice et al. recruited 14 patients with IPF, and the retention rate was 100% with no discontinuation of the D + Q intervention. Thirteen out of the 14 participants completed the planned clinical assessments. There was one serious adverse event reported, but most of the nonserious events were mild or moderate, with respiratory symptoms being the most frequent. The intervention led to significant and clinically meaningful improvements in physical function as measured by the 6-min walk distance, 4-m gait speed, and chair-stand time ($p < 0.05$). However, there were no significant changes in pulmonary function, clinical chemistries, frailty index, or reported health. The initial results from Justice et al.'s study suggests that senolytics could potentially improve physical function among IPF patients, indicating the need for further evaluation of D + Q treatment in larger randomized controlled trials.

In their open-label phase 1 pilot study, Hickson et al. (2019) gave nine participants with diabetic kidney disease (two of whom were female, with an average age of 68.7 years, an average BMI of 33.9 kg/m², and an average eGFR of 27.0 mL/min/1.73 m2) oral doses of 100 mg dasatinib and 1000 mg quercetin for three consecutive days. Before the study and 11 days after the senolytic treatment, adipose tissue, skin biopsies, and blood samples from the subjects were collected. The volume of senescent cells and SASP factors were measured. They discovered that within a period of 11 days, D + Q treatment resulted in a decrease in senescent cells in the adipose tissue. They concluded that administering senolytics as a *hit-and-run* treatment, where the drugs have elimination half-lives of less than 11 h as in the case of D + Q, leads to a significant reduction in senescent cell burden in humans. As Hickson et al. (2019, p. 454) explained:

> Although we are optimistic about the prospects for introducing senolytics and other agents that modulate fundamental ageing processes into clinical practice in the future, particularly in the near future for serious diseases for which there are currently no effective interventions, we must conclude with a note of caution. The field of senolytics is new. The first clinical trial of senolytic agents was only reported in January 2019. The findings reported here are preliminary results from an ongoing clinical trial of senolytics for treating dysfunction in patients with diabetic chronic kidney disease. Fewer than 150 subjects have been treated with these drugs in the context of clinical trials that we are aware of so far. In addition to side effects related to individual senolytic drugs known from other contexts in which those drugs have been used, there could turn out to be serious side-effects of senolytics as a class, which are not yet known. We caution against the use of senolytic agents outside the context of clinical trials until more is known about their effects and side effects.

Adverse events associated with D + Q treatment for humans and the feasibility of human trials in the future are currently unknown and were addressed by Nambiar et al. (2023). For their study, 12 participants older than 50 years with a diagnosis of IPF were enrolled in a blinded and randomized study. Participants were matched in pairs and one participant received either 3 weeks of D + Q treatment (dasatinib, 100 mg/day, and quercetin, 1250 mg/day) administered on three consecutive days per week or an equivalent matching placebo. Although the placebo group documented a lower number of overall nonserious adverse events (65 vs. 22), there were no instances of serious adverse events attributed to D + Q. The adverse events observed in the D + Q arm was predominantly typical in IPF patients or expected side effects of dasatinib. Notably, sleep disturbances and anxiety were more prevalent in the D + Q arm (four out of six participants) compared to the placebo arm (zero out of six participants). The assessment of frailty, pulmonary function, and physical capabilities before and after intermittent D + Q revealed no substantial differences between groups, although the study may have been underpowered to detect meaningful changes.

In summary, despite being a promising type of treatment, there are only a few published human trials of senotherapeutics. Regardless, there are many ongoing clinical trials registered on https://www.clinicaltrials.gov whose results in time will be published (see Appendix 5).

3.5   Future Opportunities for Senotherapeutic Research

3.5.1   Evaluating the Toxicity of Senotherapeutics

Any potential side effects of senolytic and senomorphic treatment are currently unknown. However, natural senolytic and senomorphic compounds may be more suitable for use in humans because they may have low toxicity. However, even *natural* compounds can be toxic if taken in high doses. The mechanisms of action of most natural senolytics and senomorphics are not well understood, and their molecular targets have not been identified and characterized, making it difficult to determine and improve their therapeutic potential. Most natural compounds have only been studied in cell cultures. Thus, more research is needed to confirm their activity and potential side effects in animal and human populations (Boccardi & Mecocci, 2021).

3.5.2   Defining and Classifying Ageing as a Disease

The elimination of senescent cells has the potential to be an effective strategy to improve both human healthspan and lifespan. However, it is challenging to translate these findings into actual medical treatments because in many nations biological ageing is not officially recognized as a disease by regulatory agencies. However, some researchers are challenging this situation by arguing that the *World Health Organization* should classify biological ageing as a disease (Khaltourina et al., 2020). Irrespective of these debates, about whether biological ageing is a disease, researchers are evaluating the effect of anti-ageing therapies on multiple age-associated diseases in the same study. For example, the TAME study aims to determine if metformin treatment among the elderly can delay the onset of age-associated diseases rather than targeting individual diseases (Borghesan et al., 2020).

3.5.3   Performing More Studies About Treatment Regimen and Frequency

Despite its potential to increase both lifespan and healthspan, more research is needed to determine the best treatment regimen and frequency of senomorphics and senolytics. Studies in mice suggest that administering senolytics intermittently, starting in middle age, may be the most beneficial approach. This *hit-and-run* approach would prevent the ongoing accumulation of senescent cells throughout a person's lifetime, which is a hallmark of the ageing process. At the same time, the risk of tumor development and insufficient wound healing would be minimal because senescent cells and the transitory levels of SASPs that they produce would still exist. However, it is currently unknown how long it takes for senescent cells to re-accumulate beyond a safe threshold after a cycle of clearance. It is also unknown if and what individual factors, such as genetics or sex, can impact the rate of senescent cell accumulation. Additionally, it is yet to be determined if certain types of senescent cells might develop resistance to senotherapeutics, analogous to how cancer cells can become resistant to cancer therapies (Borghesan et al. 2020).

3.5.4   Combining Senotherapies with Other Anti-Ageing Treatments

Studies suggest that removing senescent cells alone can delay age-associated diseases and disabilities and thus improve both healthspan and lifespan among mice. However, it is possible that this approach results in a partial rejuvenation effect. To fully create an anti-ageing treatment, it may be necessary to combine senotherapy with other strategies that reverse or slow the process of biological ageing, such as calorie restriction, stem cell transplantation, or tissue reprogramming. Therefore, while it is important to evaluate the safety and effectiveness of senotherapies, more research should also be done on combining senotherapeutics with other anti-ageing approaches (Borghesan et al., 2020).

3.5.5   Addressing the Heterogeneity of Senescent Cells

A major limitation in the senescence field is the lack of single, universal or model-specific biomarkers to identify senescent cells in culture or tissue samples. At present, the identification of senescent cells relies on a combination of multiple markers that, when present simultaneously, can discriminate between stably arrested senescent cells and quiescent or differentiated counterparts. (Di Micco et al., 2021, p. 76)

Currently, there is an incomplete understanding of the physical characteristics of senescent cells. Studies have suggested that senescent cells vary and that different factors, such as the type of cell, what causes the senescence, where the cell is located, and environmental conditions, can affect their production of SASP and other senescent cell features. This variation in physical characteristics suggests that different types of senescent cells may exist and not all of them may be harmful. Cataloguing different subtypes of senescent cells may help to create more efficacious and better tolerated senotherapies (Borghesan et al., 2020).

3.5.6 Validating if Senotherapies Slow Ageing

When senescent cells are eliminated, the neighboring non-senescent cells will divide to fill the empty space. However, as these cells divide, they will also experience telomere shortening, which is a factor that causes cellular senescence. Therefore, while senotherapy might temporarily relieve the negative effects caused by cellular senescence, in the long-term, they may accelerate the development of senescent cells and the tissue dysfunction that they produce (Lagoumtzi & Chondrogianni, 2021). This opinion has also been expressed by Kowald and Kirkwood (2021, p. 1), who stated:

> We perform numerical simulations of senescent cell accumulation and senolytic treatment in an ageing population. The simulations suggest that while senolytics diminish the burden of senescent cells, they may also impair the general repair capacity of the organism, leading to a faster accumulation post-treatment of new senescent cells.

3.5.7 Preventing Senotherapies from Inhibiting the Therapeutic Benefits of Senescent Cells

The SASPs that senescent cells produce provide therapeutic benefits, such as suppressing tumor growth and promoting tissue repair and wound healing. Thus, eliminating senescent cells completely would have negative health consequences (Föger-Samwald et al., 2022). To ensure that senescent cells continue to provide therapeutic benefits, it is important to conduct more research to determine the most effective schedule for administering senotherapeutics (e.g., weekly, monthly, bimonthly). Additionally, more research needs to be conducted to discover the best method to deliver senotherapeutics to specific organs (Palmer et al., 2022).

3.5.8 Examining the Long-Term Therapeutic Benefits of Senotherapies

A lot of our understanding about how senescent cells contribute to the creation of age-associated disease comes from research on animal models. Consequently, there is uncertainty about if senotherapies are safe or effective in humans, which is crucial information for developing treatments for patients. Senotherapy studies involving animals are limited to 2–3 years, which is most likely a much shorter time frame than what would be required for the use of senotherapies in humans. Therefore, it is not possible to assess possible long-term negative effects or toxicities of senescent cell elimination using current models and methods, as a longer observation period is required (Di Micco et al., 2021).

It is crucial to conduct large, randomized controlled trials to evaluate the safety, effectiveness, and target engagement of senotherapeutics to confirm the results from early-phase clinical trials. If senotherapeutics are safe and effective in patients with age-associated diseases, it may become acceptable to test the efficacy of such medications for less severe senescence-linked disorders. Such human trials could follow a strategy like that planned for metformin in the TAME study. TAME will test if metformin delays the appearance of a second age-associated disorder in patients who already have one such disorder, it will not be a trial including completely healthy older adults. If efforts to extend human healthspan are successful, studies in the future could potentially aim to evaluate the role of senotherapeutics in extending human lifespan (Chaib et al., 2022).

3.5.9   Conducting More Research About Senotherapeutics

The field of senotherapeutics is relatively novel, and consequently there are still many unanswered questions about the therapeutic benefits that senotherapy provides (see Table 3.1). As listed in the table below, several researchers have posed questions that studies in the future about senotherapy should answer.

Table 3.1 Unanswered questions about the therapeutic benefits of senotherapy

Reference	Question
Borghesan et al. (2020)	• Is it possible to target mechanisms that are unique to senescent cells to avoid side effects? • How heterogeneous is the phenotype of age-related senescence? • Can we specifically interfere with detrimental senescence? • What are the biomarkers that can be exploited for detection of senescent cells in vivo and to monitor the efficacy of senotherapies? • Will the selective elimination of senescent cells have a systemic health improvement? • How do we evaluate the effectiveness of senescent cell removal during aging? • How frequent and from which age should senotherapies be provided to achieve maximum healthspan improvement? • What are the effects of combining senotherapies with other rejuvenation strategies?

Table 3.1 (continued)

Reference	Question
Pignolo et al. (2020)	• Do senescent cells serve any beneficial functions, such as providing an inflammatory milieu during tissue healing? • What are the tissue-specific thresholds for senescent cell burden that, once exceeded, promote tissue dysfunction and systemic deleterious bystander effects? • Does the clearance of senescent cells reduce the influence of other primary ageing processes? • How infrequently can senolytic agents be given to minimize senescent cell burden? • How can combinations of senotherapeutic agents be devised to optimize senescent cell clearance? • Can immune system modulation synergize with senotherapeutic agents to promote senescent cell clearance? • Will senolytics synergize with disease-specific treatments to yield a more than additive beneficial effect, for example, in cardiac disease? • Are paracrine and other bystander effects the primary mechanism for the pathophysiological consequences of senescent cell accumulation? • Do immune clearance mechanisms have the same accessibility to different senescent cell niches (e.g., fat vs. bone vs. nevi)?
Sieben et al. (2018)	• To what extent can combination cancer therapy and senotherapy be employed to improve therapeutic efficacy, reduce risk of recurrence, and ultimately improve patient outcome? • Can removal of age-related senescent cells in humans reduce cancer risk? • Do different cell/tumor types have different dependencies on senescent cells, in other words more or less beneficial or detrimental roles, within their niche? • What is the mechanism for senescent cell induction of regenerative capacity in neighboring cells with short-term exposure, and can this contribute to the protumorigenic properties of senescent cells? • Which properties of senescent cells determine their role in immune attraction or deterrence, and how can these be differentially mediated in senescent cells induced by similar mechanisms? Does immune efficiency underlie these differences? • Do beneficial tumor-suppressing senescent cells modulate immunosurveillance differently from detrimental, cancer-promoting senescent cells? • How do senescent cell features and SASP from senescent cells induced by different mechanisms (oncogene-induced, therapy-induced, tumor-induced, age-related, and bystander-induced) differ in vivo, and how does this impact the tumor microenvironment and immune surveillance? • Which SASP components are involved in driving growth and bystander senescence in neighboring cells, immune attraction, and immune deterrence in vivo? • How do tumors/neoplastic cells induce senescence in neighboring cells/tumor stroma?

(continued)

Table 3.1 (continued)

Reference	Question
Cohn et al. (2022)	• Is there a universal signature for senescent cells across all tissues? And are there tissue-specific or disease-specific senescent cell signatures? • Are there unique signatures of senescent cell populations considered to be pathologic versus those that may be beneficial? • Is it feasible to design different senolytics based on heterogeneity to more precisely target diseases? • Do senescent cell signatures change over time? • Does senescence burden in one tissue correlate with that in other tissues within the same individual? • What are the precise senescent cell populations targeted by senolytics? • What are the unintended effects of senolytics on non-senescent cells? • What is happening in the surrounding tissue area when cells are targeted by senolytics? • What are the p16high and p21high cells targeted by various transgenic mouse models? • What is the optimal way to analyze senescent cell data?

3.6   Conclusion

Currently, the therapeutic potential of many different senomorphic and senolytic medications are being explored. Consequently, it is beyond the scope of this book to explore the literature about each of these candidates. Thus, it was decided that the literature about the senolytic potential of dasatinib and quercetin (D + Q) and the senomorphic potential of metformin and rapamycin would be canvassed. Despite their potential, our knowledge about senotherapeutics is still in its infancy, and there remains many opportunities to examine this new and emerging field of medicine. Some of these unexplored areas were presented at the end of this chapter.

Further Reading

Ellison-Hughes, G. M. (2020). First evidence that senolytics are effective at decreasing senescent cells in humans. *eBioMedicine, 56*, 102473. https://doi.org/10.1016/j.ebiom.2019.09.053

Gasek, N. S., Kuchel, G. A., Kirkland, J. L., & Xu, M. (2021). Strategies for targeting senescent cells in human disease. *Nature Aging, 1*(10), 870–879. https://doi.org/10.1038/s43587-021-00121-8

Gonzales, M. M., Krishnamurthy, S., Garbarino, V., Daeihagh, A. S., Gillispie, G. J., Deep, G., Craft, S., & Orr, M. E. (2021). A geroscience motivated approach to treat Alzheimer's disease: Senolytics move to clinical trials. *Mechanisms of Ageing and Development, 200*, 111589. https://doi.org/10.1016/j.mad.2021.111589

Konopka, A. R., & Miller, B. F. (2019). Taming expectations of metformin as a treatment to extend healthspan. *GeroScience, 41*(2), 101–108. https://doi.org/10.1007/s11357-019-00057-3

Palmer, A. K., Tchkonia, T., & Kirkland, J. L. (2021). Senolytics: Potential for alleviating diabetes and its complications. *Endocrinology, 162*(8), bqab058. https://doi.org/10.1210/endocr/bqab058

Piskovatska, V., Stefanyshyn, N., Storey, K. B., Vaiserman, A. M., & Lushchak, O. (2019). Metformin as a geroprotector: Experimental and clinical evidence. *Biogerontology, 20*(1), 33–48. https://doi.org/10.1007/s10522-018-9773-5

Sunjaya, A. P., & Sunjaya, A. F. (2021). Targeting ageing and preventing organ degeneration with metformin. *Diabetes & Metabolism, 47*(1), 101203. https://doi.org/10.1016/j.diabet.2020.09.009

Wang, C., Chen, B., Feng, Q., Nie, C., & Li, T. (2020). Clinical perspectives and concerns of metformin as an anti-aging drug. *Aging Medicine (Milton (N.S.W)), 3*(4), 266–275. https://doi.org/10.1002/agm2.12135

Zimmerman, S. C., Ferguson, E. L., Choudhary, V., Ranatunga, D. K., Oni-Orisan, A., Hayes-Larson, E., Duarte Folle, A., Mayeda, E. R., Whitmer, R. A., Gilsanz, P., Power, M. C., Schaefer, C., Glymour, M. M., & Ackley, S. F. (2023). Metformin cessation and dementia incidence. *JAMA Network Open, 6*(10), e2339723. https://doi.org/10.1001/jamanetworkopen.2023.39723

References

Boccardi, V., & Mecocci, P. (2021). Senotherapeutics: Targeting senescent cells for the main age-related diseases. *Mechanisms of Ageing and Development, 197*, 111526. https://doi.org/10.1016/j.mad.2021.111526

Boni, J. P., Leister, C., Hug, B., Burns, J., & Sonnichsen, D. (2012). A single-dose placebo- and moxifloxacin-controlled study of the effects of temsirolimus on cardiac repolarization in healthy adults. *Cancer Chemotherapy and Pharmacology, 69*(6), 1433–1442. https://doi.org/10.1007/s00280-012-1845-7

Borghesan, M., Hoogaars, W. M. H., Varela-Eirin, M., Talma, N., & Demaria, M. (2020). A senescence-centric view of aging: Implications for longevity and disease. *Trends in Cell Biology, 30*(10), 777–791. https://doi.org/10.1016/j.tcb.2020.07.002

Bruyn, G. A., Tate, G., Caeiro, F., Maldonado-Cocco, J., Westhovens, R., Tannenbaum, H., Bell, M., Forre, O., Bjorneboe, O., Tak, P. P., Abeywickrama, K. H., Bernhardt, P., van Riel, P. L., & RADD Study Group. (2008). Everolimus in patients with rheumatoid arthritis receiving concomitant methotrexate: A 3-month, double-blind, randomised, placebo-controlled, parallel-group, proof-of-concept study. *Annals of the Rheumatic Diseases, 67*(8), 1090–1095. https://doi.org/10.1136/ard.2007.078808

Campbell, J. M., Bellman, S. M., Stephenson, M. D., & Lisy, K. (2017). Metformin reduces all-cause mortality and diseases of ageing independent of its effect on diabetes control: A systematic review and meta-analysis. *Ageing Research Reviews, 40*, 31–44. https://doi.org/10.1016/j.arr.2017.08.003

Chaib, S., Tchkonia, T., & Kirkland, J. L. (2022). Cellular senescence and senolytics: The path to the clinic. *Nature Medicine, 28*(8), 1556–1568. https://doi.org/10.1038/s41591-022-01923-y

Chen, S., Gan, D., Lin, S., Zhong, Y., Chen, M., Zou, X., Shao, Z., & Xiao, G. (2022). Metformin in aging and aging-related diseases: Clinical applications and relevant mechanisms. *Theranostics, 12*(6), 2722–2740. https://doi.org/10.7150/thno.71360

Chung, C. L., Lawrence, I., Hoffman, M., Elgindi, D., Nadhan, K., Potnis, M., Jin, A., Sershon, C., Binnebose, R., Lorenzini, A., & Sell, C. (2019). Topical rapamycin reduces markers of senescence and aging in human skin: An exploratory, prospective, randomized trial. *GeroScience, 41*(6), 861–869. https://doi.org/10.1007/s11357-019-00113-y

Cohn, R. L., Gasek, N. S., Kuchel, G. A., & Xu, M. (2022). The heterogeneity of cellular senescence: Insights at the single-cell level. *Trends in Cell Biology, 33*(1), 9–17. https://doi.org/10.1016/j.tcb.2022.04.011

Di Micco, R., Krizhanovsky, V., Baker, D., & d'Adda di Fagagna, F. (2021). Cellular senescence in ageing: From mechanisms to therapeutic opportunities. *Nature Reviews. Molecular Cell Biology, 22*(2), 75–95. https://doi.org/10.1038/s41580-020-00314-w

Dickinson, J. M., Drummond, M. J., Fry, C. S., Gundermann, D. M., Walker, D. K., Timmerman, K. L., Volpi, E., & Rasmussen, B. B. (2013). Rapamycin does not affect post-absorptive protein metabolism in human skeletal muscle. *Metabolism: Clinical and Experimental, 62*(1), 144–151. https://doi.org/10.1016/j.metabol.2012.07.003

Drummond, M. J., Fry, C. S., Glynn, E. L., Dreyer, H. C., Dhanani, S., Timmerman, K. L., Volpi, E., & Rasmussen, B. B. (2009). Rapamycin administration in humans blocks the contraction-induced increase in skeletal muscle protein synthesis. *The Journal of Physiology, 587*(Pt 7), 1535–1546. https://doi.org/10.1113/jphysiol.2008.163816

Dugel, P. U., Blumenkranz, M. S., Haller, J. A., Williams, G. A., Solley, W. A., Kleinman, D. M., & Naor, J. (2012). A randomized, dose-escalation study of subconjunctival and intravitreal injections of sirolimus in patients with diabetic macular edema. *Ophthalmology, 119*(1), 124–131. https://doi.org/10.1016/j.ophtha.2011.07.034

Engelmann, C., & Tacke, F. (2022). The potential role of cellular senescence in non-alcoholic fatty liver disease. *International Journal of Molecular Sciences, 23*(2), 652. https://doi.org/10.3390/ijms23020652

Föger-Samwald, U., Kerschan-Schindl, K., Butylina, M., & Pietschmann, P. (2022). Age related osteoporosis: Targeting cellular senescence. *International Journal of Molecular Sciences, 23*(5), 2701. https://doi.org/10.3390/ijms23052701

Gensler, G., Clemons, T. E., Domalpally, A., Danis, R. P., Blodi, B., Wells, J., 3rd, Rauser, M., Hoskins, J., Hubbard, G. B., Elman, M. J., Fish, G. E., Brucker, A., Margherio, A., & Chew, E. Y. (2018). Treatment of geographic atrophy with intravitreal sirolimus: The age-related eye disease study 2 ancillary study. *Ophthalmology Retina, 2*(5), 441–450. https://doi.org/10.1016/j.oret.2017.08.015

Gundermann, D. M., Walker, D. K., Reidy, P. T., Borack, M. S., Dickinson, J. M., Volpi, E., & Rasmussen, B. B. (2014). Activation of mTORC1 signaling and protein synthesis in human muscle following blood flow restriction exercise is inhibited by rapamycin. *American Journal of Physiology. Endocrinology and Metabolism, 306*(10), E1198–E1204. https://doi.org/10.1152/ajpendo.00600.2013

Hickson, L. J., Langhi Prata, L. G. P., Bobart, S. A., Evans, T. K., Giorgadze, N., Hashmi, S. K., Herrmann, S. M., Jensen, M. D., Jia, Q., Jordan, K. L., Kellogg, T. A., Khosla, S., Koerber, D. M., Lagnado, A. B., Lawson, D. K., LeBrasseur, N. K., Lerman, L. O., McDonald, K. M., McKenzie, T. J., Passos, J. F., et al. (2019). Senolytics decrease senescent cells in humans: Preliminary report from a clinical trial of Dasatinib plus Quercetin in individuals with diabetic kidney disease. *eBioMedicine, 47*, 446–456. https://doi.org/10.1016/j.ebiom.2019.08.069

Hörbelt, T., Kahl, A. L., Kolbe, F., Hetze, S., Wilde, B., Witzke, O., & Schedlowski, M. (2020). Dose-dependent acute effects of everolimus administration on immunological, neuroendocrine and psychological parameters in healthy men. *Clinical and Translational Science, 13*(6), 1251–1259. https://doi.org/10.1111/cts.12812

Huang, Y., Wang, B., Hassounah, F., Price, S. R., Klein, J., Mohamed, T. M. A., Wang, Y., Park, J., Cai, H., Zhang, X., & Wang, X. H. (2023). The impact of senescence on muscle wasting in chronic kidney disease. *Journal of Cachexia, Sarcopenia and Muscle, 14*(1), 126–141. https://doi.org/10.1002/jcsm.13112

Islam, M. T., Tuday, E., Allen, S., Kim, J., Trott, D. W., Holland, W. L., Donato, A. J., & Lesniewski, L. A. (2023). Senolytic drugs, dasatinib and quercetin, attenuate adipose tissue inflammation, and ameliorate metabolic function in old age. *Aging Cell, 22*(2), e13767. https://doi.org/10.1111/acel.13767

Justice, J. N., Nambiar, A. M., Tchkonia, T., LeBrasseur, N. K., Pascual, R., Hashmi, S. K., Prata, L., Masternak, M. M., Kritchevsky, S. B., Musi, N., & Kirkland, J. L. (2019). Senolytics in idiopathic pulmonary fibrosis: Results from a first-in-human, open-label, pilot study. *eBioMedicine, 40*, 554–563. https://doi.org/10.1016/j.ebiom.2018.12.052

Khaltourina, D., Matveyev, Y., Alekseev, A., Cortese, F., & Ioviţă, A. (2020). Aging fits the disease criteria of the international classification of diseases. *Mechanisms of Ageing and Development, 189*, 111230. https://doi.org/10.1016/j.mad.2020.111230

Kowald, A., & Kirkwood, T. B. L. (2021). Senolytics and the compression of late-life mortality. *Experimental Gerontology, 155*, 111588. https://doi.org/10.1016/j.exger.2021.111588

Kraig, E., Linehan, L. A., Liang, H., Romo, T. Q., Liu, Q., Wu, Y., Benavides, A. D., Curiel, T. J., Javors, M. A., Musi, N., Chiodo, L., Koek, W., Gelfond, J. A. L., & Kellogg, D. L., Jr. (2018). A randomized control trial to establish the feasibility and safety of rapamycin treatment in an older human cohort: Immunological, physical performance, and cognitive effects. *Experimental Gerontology, 105*, 53–69. https://doi.org/10.1016/j.exger.2017.12.026

Krebs, M., Brunmair, B., Brehm, A., Artwohl, M., Szendroedi, J., Nowotny, P., Roth, E., Fürnsinn, C., Promintzer, M., Anderwald, C., Bischof, M., & Roden, M. (2007). The Mammalian target of rapamycin pathway regulates nutrient-sensitive glucose uptake in man. *Diabetes, 56*(6), 1600–1607. https://doi.org/10.2337/db06-1016

Krzystyniak, A., Wesierska, M., Petrazzo, G., Gadecka, A., Dudkowska, M., Bielak-Zmijewska, A., Mosieniak, G., Figiel, I., Wlodarczyk, J., & Sikora, E. (2022). Combination of dasatinib and quercetin improves cognitive abilities in aged male Wistar rats, alleviates inflammation and changes hippocampal synaptic plasticity and histone H3 methylation profile. *Aging, 14*(2), 572–595. https://doi.org/10.18632/aging.203835

Lagoumtzi, S. M., & Chondrogianni, N. (2021). Senolytics and senomorphics: Natural and synthetic therapeutics in the treatment of aging and chronic diseases. *Free Radical Biology & Medicine, 171*, 169–190. https://doi.org/10.1016/j.freeradbiomed.2021.05.003

Lee, D. J. W., Hodzic Kuerec, A., & Maier, A. B. (2024). Targeting ageing with rapamycin and its derivatives in humans: A systematic review. *The Lancet. Healthy Longevity, 5*(2), e152–e162. https://doi.org/10.1016/S2666-7568(23)00258-1

Mannick, J. B., Del Giudice, G., Lattanzi, M., Valiante, N. M., Praestgaard, J., Huang, B., Lonetto, M. A., Maecker, H. T., Kovarik, J., Carson, S., Glass, D. J., & Klickstein, L. B. (2014). mTOR inhibition improves immune function in the elderly. *Science Translational Medicine, 6*(268), 268ra179. https://doi.org/10.1126/scitranslmed.3009892

Mannick, J. B., Teo, G., Bernardo, P., Quinn, D., Russell, K., Klickstein, L., Marshall, W., & Shergill, S. (2021). Targeting the biology of ageing with mTOR inhibitors to improve immune function in older adults: Phase 2b and phase 3 randomised trials. *The Lancet. Healthy Longevity, 2*(5), e250–e262. https://doi.org/10.1016/S2666-7568(21)00062-3

Minturn, R. J., Bracha, P., Klein, M. J., Chhablani, J., Harless, A. M., & Maturi, R. K. (2021). Intravitreal sirolimus for persistent, exudative age-related macular degeneration: A pilot study. *International Journal of Retina and Vitreous, 7*, 1–10. https://doi.org/10.1186/s40942-021-00281-0

Mohammed, I., Hollenberg, M. D., Ding, H., & Triggle, C. R. (2021). A critical review of the evidence that metformin is a putative anti-aging drug that enhances healthspan and extends lifespan. *Frontiers in Endocrinology, 12*, 718942. https://doi.org/10.3389/fendo.2021.718942

Nambiar, A., Kellogg, D., 3rd, Justice, J., Goros, M., Gelfond, J., Pascual, R., Hashmi, S., Masternak, M., Prata, L., LeBrasseur, N., Limper, A., Kritchevsky, S., Musi, N., Tchkonia, T., & Kirkland, J. (2023). Senolytics dasatinib and quercetin in idiopathic pulmonary fibrosis: Results of a phase I, single-blind, single-center, randomized, placebo-controlled pilot trial on feasibility and tolerability. *eBioMedicine, 90*, 104481. https://doi.org/10.1016/j.ebiom.2023.104481

Novais, E. J., Tran, V. A., Johnston, S. N., Darris, K. R., Roupas, A. J., Sessions, G. A., Shapiro, I. M., Diekman, B. O., & Risbud, M. V. (2021). Long-term treatment with senolytic drugs Dasatinib and Quercetin ameliorates age-dependent intervertebral disc degeneration in mice. *Nature Communications, 12*(1), 5213. https://doi.org/10.1038/s41467-021-25453-2

Nussenblatt, R. B., Byrnes, G., Sen, H. N., Yeh, S., Faia, L., Meyerle, C., Wroblewski, K., Li, Z., Liu, B., Chew, E., Sherry, P. R., Friedman, P., Gill, F., & Ferris, F., 3rd. (2010). A randomized pilot study of systemic immunosuppression in the treatment of age-related macular degeneration with choroidal neovascularization. *Retina (Philadelphia, Pa.), 30*(10), 1579–1587. https://doi.org/10.1097/IAE.0b013e3181e7978e

Palma, J. A., Martinez, J., Millar Vernetti, P., Ma, T., Perez, M. A., Zhong, J., Qian, Y., Dutta, S., Maina, K. N., Siddique, I., Bitan, G., Ades-Aron, B., Shepherd, T. M., Kang, U. J., & Kaufmann, H. (2022). mTOR inhibition with sirolimus in multiple system atrophy: A randomized, double-blind, placebo-controlled futility trial and 1-year biomarker longitudinal analysis. *Movement Disorders: Official Journal of the Movement Disorder Society, 37*(4), 778–789. https://doi.org/10.1002/mds.28923

Palmer, A. K., Tchkonia, T., & Kirkland, J. L. (2022). Targeting cellular senescence in metabolic disease. *Molecular Metabolism, 66*, 101601. https://doi.org/10.1016/j.molmet.2022.101601

Petrou, P. A., Cunningham, D., Shimel, K., Harrington, M., Hammel, K., Cukras, C. A., Ferris, F. L., Chew, E. Y., & Wong, W. T. (2014). Intravitreal sirolimus for the treatment of geographic atrophy: Results of a phase I/II clinical trial. *Investigative Ophthalmology & Visual Science, 56*(1), 330–338. https://doi.org/10.1167/iovs.14-15877

Pignolo, R. J., Passos, J. F., Khosla, S., Tchkonia, T., & Kirkland, J. L. (2020). Reducing senescent cell burden in aging and disease. *Trends in Molecular Medicine, 26*(7), 630–638. https://doi.org/10.1016/j.molmed.2020.03.005

Raffaele, M., Kovacovicova, K., Frohlich, J., Lo Re, O., Giallongo, S., Oben, J. A., Faldyna, M., Leva, L., Giannone, A. G., Cabibi, D., & Vinciguerra, M. (2021). Mild exacerbation of obesity- and age-dependent liver disease progression by senolytic cocktail dasatinib + quercetin. *Cell Communication and Signaling, 19*(1), 44. https://doi.org/10.1186/s12964-021-00731-0

Salerno, N., Marino, F., Scalise, M., Salerno, L., Molinaro, C., Filardo, A., Chiefalo, A., Panuccio, G., De Angelis, A., Urbanek, K., Torella, D., & Cianflone, E. (2022). Pharmacological clearance of senescent cells improves cardiac remodeling and function after myocardial infarction in female aged mice. *Mechanisms of Ageing and Development, 208*, 111740. https://doi.org/10.1016/j.mad.2022.111740

Seyfarth, H. J., Hammerschmidt, S., Halank, M., Neuhaus, P., & Wirtz, H. R. (2013). Everolimus in patients with severe pulmonary hypertension: A safety and efficacy pilot trial. *Pulmonary Circulation, 3*(3), 632–638. https://doi.org/10.1086/674311

Sieben, C. J., Sturmlechner, I., van de Sluis, B., & van Deursen, J. M. (2018). Two-step senescence-focused cancer therapies. *Trends in Cell Biology, 28*(9), 723–737. https://doi.org/10.1016/j.tcb.2018.04.006

Sierra-Ramirez, A., López-Aceituno, J. L., Costa-Machado, L. F., Plaza, A., Barradas, M., & Fernandez-Marcos, P. J. (2020). Transient metabolic improvement in obese mice treated with navitoclax or dasatinib/quercetin. *Aging, 12*(12), 11337–11348. https://doi.org/10.18632/aging.103607

Triggle, C. R., Mohammed, I., Bshesh, K., Marei, I., Ye, K., Ding, H., MacDonald, R., Hollenberg, M. D., & Hill, M. A. (2022). Metformin: Is it a drug for all reasons and diseases? *Metabolism: Clinical and Experimental, 133*, 155223. https://doi.org/10.1016/j.metabol.2022.155223

Wen, H. Y., Wang, J., Zhang, S. X., Luo, J., Zhao, X. C., Zhang, C., et al. (2019). Low-dose sirolimus immunoregulation therapy in patients with active rheumatoid arthritis: A 24-week follow-up of the randomized, open-label, parallel-controlled trial. *Journal of Immunology Research, 2019*. https://doi.org/10.1155/2019/7684352

Chapter 4
Final Comments

Abstract This chapter summarizes the literature examined in this book and explains recent developments with the World Health Organization classifying human ageing as a biological factor that causes diseases. It also reiterates the notion that senotherapy may be an effective strategy for improving human lifespan and healthspan.

Keywords Age-associated diseases · Ageing · Healthspan · Lifespan · Senotherapy · World Health Organization (WHO)

The research presented in the previous two chapters strongly suggests that the continual accumulation of senescent cells as biological age progresses is the root cause of many age-associated diseases that occur among the elderly. If this suggestion is correct, it lends convincing support to the *geroscience hypothesis*, which suggests that since biological ageing is the underlying cause of most age-associated diseases, interventions that can slow or reverse the process of biological ageing would also prevent, delay, or alleviate these age-associated diseases (Pignolo et al., 2020). Thus, treating biological ageing as a disease is being contemplated as a viable solution to either prevent or cure many age-associated diseases that occur among the elderly. Recently, the World Health Organization (WHO) controversially incorporated into the 11th version of the *International Classification of Diseases* (ICD-11) an extension code for *Ageing-related* (XT9T) diseases, which are defined as those "caused by pathological processes which persistently lead to the loss of organism's adaptation and progress in older ages" (The Lancet Diabetes Endocrinology, 2018). The potential consequences of incorporating this code were articulated in the *Lancet Diabetes Endocrinology*:

> *Although implementation of the extension code XT9T in ICD-11 is not tantamount to formal recognition of ageing as a disease, it does signal acknowledgment by WHO of ageing as a major disease risk factor and of the considerable public health problem posed by ageing-related diseases. Whether this move will be enough to entice the pharmaceutical industry to initiate development programmes targeting the ageing process and, more broadly, human lifespan remains to be seen. However, given the impending economic threat posed by*

G. Bennett, *Senotherapy*, SpringerBriefs in Modern Perspectives on Disability Research, https://doi.org/10.1007/978-981-97-3637-9_4

population ageing and the potential benefits of intervention, ranging from better health in old age for individuals to decreased health-care costs, increased workplace productivity, and increased economic growth for countries, progress on this front is vital. (The Lancet Diabetes Endocrinology, 2018, p. 587)

Since there are many potential strategies that are being explored to slow or reverse biological ageing, it was not possible to review the literature about each strategy. Instead, the literature about senotherapeutic medications designed to either remove senescent cells (i.e., senolytics) or inhibit their detrimental effects (i.e., senomorphics) was explored. As shown in Chap. 3, there is a considerable amount of animal-based research measuring the therapeutic potential of senotherapeutics. However, as of 2024, there were only a few human trials of senotherapeutics (Hickson et al., 2019; Justice et al., 2019; Nambiar et al., 2023). If these and other trials yield promising results, it might be possible to either eradicate age-associated diseases or reduce their severity among the elderly. As Burton and Faragher (2018, p. 456) once explained:

> *If many different diseases have a common underlying cause, such as cell senescence, it may one day be possible to treat many different diseases with a single drug. However, rather than waiting for a disease to present itself, intermittent removal of senescent cells throughout our lifetime could be utilised as a preventative measure, ensuring long and healthy lives.*

Despite their therapeutic potential, there remains many unanswered questions about senotherapies and senescent cells in general (Borghesan et al., 2020; Cohn et al., 2022; Pignolo et al., 2020; Sieben et al., 2018). Furthermore, in the quest to reduce senescent cell accumulation, it is important to ensure that the most therapeutic number of senescent cells are struck. If such cells are noticeably below a safe therapeutic threshold, then wound repair and tumour suppression would be jeopardized. Alternatively, if such cells are noticeably above this threshold, then the prospect of age-associated diseases developing increases (Chaib et al., 2022; Föger-Samwald et al., 2022). However, it has been proposed that senolytic medications could be delivered in a *hit-and-run* approach. This approach would allow the number of senescent cells to stay at a threshold where their benefits would outweigh their risks. Furthermore, a *hit-and-run* approach would maximize the prospect of senolytic medications being effective for longer since any decrease of efficacy caused by continuous usage would be minimized. Finally, perhaps one of the most profound issues that will occur with the widespread usage of senotherapy, which will need to be researched, is the impact of society adjusting its expectations about retirement and what it will mean to be elderly and not be burdened with age-associated diseases.

References

Borghesan, M., Hoogaars, W. M. H., Varela-Eirin, M., Talma, N., & Demaria, M. (2020). A senescence-centric view of aging: Implications for longevity and disease. *Trends in Cell Biology, 30*(10), 777–791. https://doi.org/10.1016/j.tcb.2020.07.002

Burton, D. G. A., & Faragher, R. G. A. (2018). Obesity and type-2 diabetes as inducers of prema-ture cellular senescence and ageing. *Biogerontology, 19*(6), 447–459. https://doi.org/10.1007/s10522-018-9763-7

Chaib, S., Tchkonia, T., & Kirkland, J. L. (2022). Cellular senescence and senolytics: The path to the clinic. *Nature Medicine, 28*(8), 1556–1568. https://doi.org/10.1038/s41591-022-01923-y

Cohn, R. L., Gasek, N. S., Kuchel, G. A., & Xu, M. (2022). The heterogeneity of cellular senescence: Insights at the single-cell level. *Trends in Cell Biology, 33*(1), 9–17. https://doi.org/10.1016/j.tcb.2022.04.011

Föger-Samwald, U., Kerschan-Schindl, K., Butylina, M., & Pietschmann, P. (2022). Age related osteoporosis: Targeting cellular senescence. *International Journal of Molecular Sciences, 23*(5), 2701. https://doi.org/10.3390/ijms23052701

Hickson, L. J., Langhi Prata, L. G. P., Bobart, S. A., Evans, T. K., Giorgadze, N., Hashmi, S. K., Herrmann, S. M., Jensen, M. D., Jia, Q., Jordan, K. L., Kellogg, T. A., Khosla, S., Koerber, D. M., Lagnado, A. B., Lawson, D. K., LeBrasseur, N. K., Lerman, L. O., McDonald, K. M., McKenzie, T. J., Passos, J. F., et al. (2019). Senolytics decrease senescent cells in humans: Preliminary report from a clinical trial of Dasatinib plus Quercetin in individuals with diabetic kidney disease. *eBioMedicine, 47*, 446–456. https://doi.org/10.1016/j.ebiom.2019.08.069

Justice, J. N., Nambiar, A. M., Tchkonia, T., LeBrasseur, N. K., Pascual, R., Hashmi, S. K., Prata, L., Masternak, M. M., Kritchevsky, S. B., Musi, N., & Kirkland, J. L. (2019). Senolytics in idiopathic pulmonary fibrosis: Results from a first-in-human, open-label, pilot study. *eBio-Medicine, 40*, 554–563. https://doi.org/10.1016/j.ebiom.2018.12.052

Nambiar, A., Kellogg, D., 3rd, Justice, J., Goros, M., Gelfond, J., Pascual, R., Hashmi, S., Masternak, M., Prata, L., LeBrasseur, N., Limper, A., Kritchevsky, S., Musi, N., Tchkonia, T., & Kirkland, J. (2023). Senolytics dasatinib and quercetin in idiopathic pulmonary fibro-sis: Results of a phase I, single-blind, single-center, randomized, placebo-controlled pilot trial on feasibility and tolerability. *eBioMedicine, 90*, 104481. https://doi.org/10.1016/j.ebiom.2023.104481

Pignolo, R. J., Passos, J. F., Khosla, S., Tchkonia, T., & Kirkland, J. L. (2020). Reducing senescent cell burden in aging and disease. *Trends in Molecular Medicine, 26*(7), 630–638. https://doi.org/10.1016/j.molmed.2020.03.005

Sieben, C. J., Sturmlechner, I., van de Sluis, B., & van Deursen, J. M. (2018). Two-step senescence-focused cancer therapies. *Trends in Cell Biology, 28*(9), 723–737. https://doi.org/10.1016/j.tcb.2018.04.006

The Lancet Diabetes Endocrinology. (2018). Opening the door to treating ageing as a disease. *The Lancet. Diabetes & endocrinology, 6*(8), 587. https://doi.org/10.1016/S2213-8587(18)30214-6

Appendices

Appendix 1: History of the Discovery and Development of Metformin to Treat Diabetes

Chronology	Brief description of observation
~1600s: Use of herbs in folklore medicine in Medieval Europe and described in *Culpeper's Complete Herbal* of 1653	Extracts of leaves and seed pods from the perennial herb, French lilac (*Galega officinalis*, also known as Italian fitch, goat's rue, Spanish sainfoin, and Professor weed) used to treat diabetes as detected as "sweet urine" and polyuria. Also used as a galactogogue in cows and goats and a variety of other maladies Later determined that active chemical was the guanidine or galegine
1844–1861 and 1878–1879	1. German chemist Adolph Strecker first described the chemical synthesis of guanidine 2. The synthesis of biguanides was carried out by German chemist Bernhard Rathke
1918: Guanidine hydrochloride	Glucose-lowering effects of guanidine observed when injected into rabbits
1922: The synthesis of the biguanide dimethyl guanidine (metformin) first described	Synthesis based on a previous description of producing guanidine thiocyanate from ammonium thiocyanate and dicyanodiamide

© The Author(s), under exclusive license to Springer Nature Singapore Pte
Ltd. 2024
G. Bennett, *Senotherapy*, SpringerBriefs in Modern Perspectives on Disability
Research, https://doi.org/10.1007/978-981-97-3637-9

Chronology	Brief description of observation
1926–1928: Description and antihyperglycemic properties of Synthalin A and B	The link between the ability of guanidine to lower blood glucose and toxicity stimulated the search for guanidines with high antihyperglycemic potency and reduced toxicity. Frank, Nothmann, and Wagner and also Graham and Linder described the effectiveness of Synthalin (two guanidine groups linked by an aliphatic chain consisting of 10 links) as a promising molecule for the treatment of diabetes. Synthalin (later renamed Synthalin A) was marketed by Schering AG and less toxic than guanidine. Synthalin B was developed with a longer aliphatic chain with 12 links and claimed to be safer. An accumulation of liver and renal toxicity reports resulted in the withdrawal of Synthalin B from the market in the 1930s and finally in Germany in the mid-1940s
1929: Metformin lowers blood glucose	Metformin injected into rabbits lowers blood glucose and determined to be the most potent of a series of compounds tested. Lack of follow-up may be linked to the discovery of insulin in 1922
1948: Approval of proguanil (chloroguanide) by the FDA to treat malaria and marketed as Paludrine	Proguanil, a structural analogue of metformin, is a prodrug that is metabolized by CYP2C19 to the active cycloguanil. Metformin was also tested in the 1940s for use in malaria, and interest recently focused on using metformin as an adjunct in combination with antimalarial drugs
1950: Metformin used to treat influenza	Metformin under the name flumamine
1957: Metformin used in humans with diabetes	Jean Sterne described the effectiveness of metformin in patients with diabetes. However, the more potent phenformin and buformin were preferred until their withdrawal from most markets in the 1970s due to the risk of lactic acidosis
1958: Toxicity study of phenformin versus Synthalin B	Comparison of liver toxicity in guinea pigs comparing Synthalin B with DBI (phenformin) Conclusion: Phenformin is a safer drug than Synthalin B
1978: Phenformin, the phenethylbiguanide relative of metformin, withdrawn from most markets	Due to an increasing number of reports of lactic acidosis and resultant high mortality, the FDA announced the withdrawal of phenformin on November 15, 1978

Chronology	Brief description of observation
1998: UKPDS (United Kingdom Prospective Diabetes Study). A landmark randomized, multicenter trial involving 23 sites and 5102 patients with newly diagnosed type 2 diabetes. UKPDS comparing insulin, sulfonylureas, and metformin	UKPDS involved 5102 patients with newly diagnosed type 2 diabetes. The study, published in 1998, reported the cardiovascular benefits of the use of metformin for diabetes. In the UKPDS 34 subgroup, 1704 overweight patients with T2DM were assigned to one of three arms: 1. Conventional therapy with diet alone 2. Intensive therapy with metformin 3. Intensive therapy with first generation, sulfonylurea chlorpropamide, and second generation, glibenclamide or insulin) The results demonstrated a reduction in diabetes-related complications and all-cause mortality for those in the metformin arm of the study compared to the other two arms of the study. Benefits were maintained after an additional 10 years of follow-up
1995: The FDA approved metformin for the treatment of type 2 diabetes	
2020: Despite the availability of many new drugs, and also formulations of insulin available, metformin maintains the position as the first-choice drug for most patients diagnosed with T2D	As concluded in a 2020 review article: *"Until further safety data becomes available for SGLT2i and GLP-1RA use in treatment-naïve individuals, we recommend that not only the efficacy but also the cost and the long-term safety profile should guide decisions in clinical practice and metformin should continue to be used as a first-line therapy for newly diagnosed individuals with T2D. The key message is to avoid therapeutic inertia, as the uptake of these 'newer' GLTs (glucose-lowering therapies) with proven cardiovascular benefits remains generally low and to consider early addition of these agents to baseline metformin therapy when indicated."* In addition: *"Metformin prescribing peaked from 55.4% in 2000 to 83.6% in 2013 among all individuals with T2D who were on at least one medication for their diabetes management in the UK (Sharma et al, 2016). Similarly, in the USA use for metformin increased from 60% in 2005 to 77% in 2016."* (Montvida et al., 2018)

Source: Triggle, C. R., Mohammed, I., Bshesh, K., Marei, I., Ye, K., Ding, H., MacDonald, R., Hollenberg, M. D., & Hill, M. A. (2022). Metformin: Is it a drug for all reasons and diseases?. *Metabolism: Clinical and Experimental, 133*, 155223. https://doi.org/10.1016/j.metabol.2022.155223

Appendix 2: Summary of Key Targets and Pathways Impacted by Metformin as Evidenced by Their Modulation in Cell Lines, *C. elegans*, *Drosophila*, and Rodents

Effects of metformin on key targets and pathways involved in regulating each hallmark of ageing

Attenuation of hallmarks of ageing	Cell lines	*C. elegans*	*Drosophila*	Rodents
Improved nutrient signaling	AMPK ↑ (Hawley et al., 2002) SIRT1 ↑ in low NAD+ concentrations (Cuyàs et al., 2018b) Insulin/IGF-1 signaling ↓ (Sarfstein et al., 2013) AGEs ↓ (Chung et al., 2017) mTORC1 ↓ via Rag-GTPase ↓ (Kalender et al., 2010), via TSC2 ↑ (Dowling et al., 2007) & via REDD1 ↑ (Ben Sahra et al., 2011)	AMPK ↑ (Onken & Driscoll, 2010, Cabreiro et al., 2013) LKB1 ↑ (Onken & Driscoll, 2010) SKN1 ↑ (Onken & Driscoll, 2010) TORC1 ↓ (Chen et al., 2017a)	AMPK ↑ (no change in lifespan) (Slack et al., 2012) Lipid stores ↓ (Slack et al., 2012)	AMPK ↑ (Martin-Montalvo et al., 2013, Howell et al., 2017) mTORC1 ↓ (Howell et al., 2017)
Enhanced intercellular communication	IKKα/β ↓, IL-6 ↓, IL-1β ↓, CXCL1/2 ↓ (Cameron et al., 2016) NF-κB pathway ↓ (Moiseeva et al., 2013) MMP-9 ↓ and STAT3 ↓ (Vasamsetti et al., 2015) Activator protein 1 (AP-1) transcription factor network ↑ and cytokine-cytokine receptor interactions modulated (Gillespie et al., 2019)	Methionine restriction (Cabreiro et al., 2013) Branched-chain amino acids ↓ (De Haes et al., 2014)	–	•IL-6 ↓, IL-1β ↓•*Lactobacilli* in the upper small intestine ↑ (Bauer et al., 2018) •*Bacteroides fragilis* ↓ and FXR signaling ↓ (Sun et al., 2018)
Ameliorated proteostasis	LAMP-1 ↑ & Beclin-1 ↑ (Wang et al., 2017) LC3 ↑ and p-ULK ↑ (Wang et al., 2018) CEBPD-mediated autophagy ↑ (Tsai et al., 2017)	HLH-30 translocalized to the nucleus with autophagy gene expression ↑ (Chen et al., 2017b)	–	α-Synuclein ↓ (Lu et al., 2016) LC3-II ↑ (enhanced autophagy) (Yan et al., 2017)

Protection against genomic instability	Ataxia-telangiectasia mutated (ATM)-dependent DNA damage response ↑ (Vazquez-Martin et al., 2011) DNA damage ↓ (Algire et al., 2012) Micronuclei ↓, dicentrics, and acentric fragments ↓, nucleoplasmic bridges ↓, rings ↓ (Cheki et al., 2016) DNA repair ↑ and DNA damage ↓ (Lee et al., 2016)	–	DNA damage ↓ and oxidative stress ↓ (Na et al., 2013)	Micronuclei ↓ (Sant' Anna et al., 2013, Cheki et al., 2021)
Regulated mitochondrial function	Mt-complex I ↓, and oxidative phosphorylation ↓, and endogenous ROS ↓ (El-Mir et al., 2000, Algire et al., 2012) mGP_{DH} ↓ and hepatic glucose production ↓ (Madiraju et al., 2014) PGC-1α ↑ (Aatsinki et al., 2014)	Mt-complex I ↓ (De Haes et al., 2014, Wu et al., 2016) PRDX-2 ↑ leading to mitohormesis (De Haes et al., 2014)	–	Mt-complex I ↑ (Wheaton et al., 2014) Energy state ↓ and AMP-ATP ↑ (Foretz et al., 2010) PGC-1α ↑ (Foretz et al., 2010)
Increased stem cell rejuvenation capacity	Gpx7-Nrf2 ↑ leading to increased lifespan in mesenchymal stem cells (Fang et al., 2018) Rejuvenation ↑, myelination ↑ in aged oligodendrocyte progenitor cells (Neumann et al., 2019) TAp73 ↑, AMPK-mediated αPKC-CBP pathway ↑, and rejuvenation of stem cells (Fatt et al., 2015)	–	Intestinal stem cell ageing ↓ mediated by Atg6 (Beclin-1) (Na et al., 2018)	Expansion of neural stem cell pool (Ruddy et al., 2019) Adult neural precursors (Fatt et al., 2015) Quiescence ↑ and delayed differentiation of satellite cells (Pavlidou et al., 2019)

Effects of metformin on key targets and pathways involved in regulating each hallmark of ageing				
Attenuation of hallmarks of ageing	Cell lines	C. elegans	Drosophila	Rodents
Regulated epigenetic alterations	Histone methyltransferase ↑ and histone acetyltransferase ↑, class II histone deacetylase ↓ and class III histone deacetylase ↑ (Bridgeman et al., 2018) Global H3K27me3 ↑ (Cuyàs et al., 2018a) Increased global DNA methylation (Cuyàs et al., 2018a) via the H19/S-adenosylhomocysteine hydrolase (SAHH) axis (Zhong et al., 2017)	S-adenosylmethionine ↓ and S-adenosylhomocysteine ↑ thereby modulating histone methylation (Cabreiro et al., 2013)	–	H3K27me3 ↑ via directly targeting KDM6A/UTX (Cuyàs et al., 2018a) DICER-1 ↑ and increased regulation of miRNA expression (Noren Hooten et al., 2016) Regulation of hydroxymethylation via AMPK/Tet2/BDNF pathway (Wang et al., 2019)
Minimized telomere attrition	TERRA and telomeric transcriptional regulation ↑ via Nrf1 and PGC-1α (Diman et al., 2016)	–	–	–
Attenuated cellular senescence	Senescence ↓ and SASP ↓ via Nrf2-mediated Gpx7 ↑ (Fang et al., 2018) SASP ↓ via anti-inflammatory properties such as NF-κB pathway ↓ (Moiseeva et al., 2013, Śmieszek et al., 2017)	–	–	p16 ↓ and p21 ↓, IL-6 ↓ and IL-8 ↓ via DICER ↑ (Noren Hooten et al., 2016)

Source: Kulkarni et al. (2020), pp. 18–19
Abbreviations: ↓ = decrease, ↑ = increase

Appendix 3: Clinical Trials Using Metformin for Targeting Biological Ageing on https://www.clinicaltrials.gov as of February 18, 2024

Study characteristic	Description
NCT number	NCT03309007
Status	Completed
Official title	A Double-Blind, Placebo-Controlled Trial of Anti-ageing, Pro-Autophagy Effects of Metformin in Adults With Prediabetes
Study start date	September 1, 2017
Study end date	August 20, 2020 (final data collection date for primary outcome measure)
Study description	The goal of this pilot and feasibility study is to investigate the effects of a short course of metformin therapy on a surrogate marker of cellular senescence and autophagy among adult patients with prediabetes. The overall hypothesis is that metformin will have beneficial effects on longevity and quality of life by inducing autophagy downstream of activating adenosine monophosphate-activated protein kinase (AMPK) and inhibiting mechanistic target of rapamycin (mTOR) through potential effects of reduced inflammation, reduced degeneration of muscle and tendon tissue, antineoplastic effects, reduced obesity and hyperglycemia, preserved cardiovascular functions, and/or the prevention of neurodegeneration (such as age-associated dementia). This pilot study will address the following aim: Demonstrate that metformin therapy will increase cellular autophagy as an inverse correlate of ageing as measured by increases in microtubule-associated protein 1A/1B-light chain 3 (LC3) scores Hypothesis 1: In addition to beneficial effects on glycemia, body weight, and body composition, metformin therapy exerts beneficial effects on surrogate measures of autophagy and ageing Primary outcome: Increased levels of LC3 in leukocytes
Condition	• Prediabetes • Ageing

Study arms	• Experimental: Metformin
	    Metformin started at 500 mg po twice daily (BID) and then titrated up to 1000 mg po q morning (AM) and 500 po q evening (PM) over the course of 1 month, as tolerated
	        Intervention: Drug—Metformin
	• Placebo comparator: Placebo oral tablet
	    Near-identical CaCO3 as a placebo oral tablet will be started at 648 mg po BID and then titrated up to 1296 mg po q AM and 648 mg po q PM over the course of 1 month, as tolerated
	        Intervention: Drug—placebo oral tablet

Actual enrollment	24
Sex/gender	All sexes eligible for study
Ages	30–70 years (adult, older adult)
NCT number	NCT02432287
Status	Completed
Official title	Metformin in Longevity Study (MILES)
Study start date	October 2014
Study end date	September 2015 (final data collection date for primary outcome measure)

Study characteristic	Description
Study description	Ageing in humans is a well-established primary risk factor for many disabling diseases and conditions, among them diabetes, cardiovascular disease, Alzheimer's disease, and cancer. In fact, the risk of death from these causes is dramatically accelerated (100–1000-fold) between the ages of 35 and 85 years. For this reason, there is a need for the development of new interventions to improve and maintain health into old age—to improve "healthspan"
	Several mechanisms have been shown to delay the ageing process, resulting in improved healthspan in animal models, including mammals. These include caloric restriction, alteration in GH/IGF1 pathways, and use of several drugs such as resveratrol (SIRT1 activator) and rapamycin (mTOR inhibitor). At Einstein, the investigators have been working to discover pathways associated with exceptional longevity. The investigators propose the study of drugs already in common clinical use (and FDA approved) for a possible alternative purpose—healthy ageing. The investigators' goal is to identify additional mechanisms involved in ageing, the delay of ageing, and the prevention of age-related diseases. In this proposal, the investigators explore the possibility of a commonly used drug, metformin, to reverse relevant aspects of the physiology and biology of ageing
	Metformin is an FDA-approved drug in common use in the US since the 1990s. It is the first-line drug of choice for prevention and treatment of type 2 diabetes (T2DM). The effect of metformin on ageing has been extensively studied and has been associated with longevity in many rodent models. Metformin also extends the lifespan of nematodes, suggesting an evolutionarily conserved mechanism. A recent high impact study demonstrated that metformin reduces oxidative stress and inflammation and extends both lifespan and health span in a mouse model
	If indeed metformin is an "anti-ageing" drug, its administration should be associated with less age-related disease in general, rather than the decreased incidence of a single age-related disease. This notion led investigators to further study whether anti-ageing effects can be demonstrated in the type 2 diabetes population. Notably, in the United Kingdom Prospective Diabetes Study (UKPDS), metformin, compared with other antidiabetic drugs, demonstrated a decreased risk of cardiovascular disease. This has been suggested in other studies and meta-analyses and remains an active area of research
	In addition, numerous epidemiologic studies have shown an association of metformin use with a decreased risk of cancer, as well as decreased cancer mortality. There is also evidence from studies performed both in vitro and in vivo of metformin's role in attenuating tumorigenesis. The mechanisms proposed relate to its effects on reducing insulin levels, improving insulin action, decreasing IGF-1 signaling (central to mammalian longevity), activating AMP kinase. In fact, metformin's potential protective effect against cancer has been gaining much attention, with over 100 ongoing studies registered on the Clinical Trials.gov website
	To characterize pathways associated with increased lifespan and healthspan, the investigators plan to compile a repository of muscle and adipose biopsy samples obtained from young healthy subjects and older adults before and after treatment with potential anti-ageing drugs. RNA-Seq analysis will be used to identify a unique biological "fingerprint" for ageing in these tissues by comparing changes in gene expression in older adults postdrug therapy to the profiles of young healthy subjects. This overall approach is supported by a grant from the Glenn Foundation for the Study of the Biology of Human Aging
	The investigators believe that if metformin changes the biology of ageing in tissues to a younger profile, it supports the notion that this drug may have more widespread use—as an "anti-ageing" drug

Study characteristic	Description
Condition	Ageing

Study arms	• Experimental: Metformin
	    Metformin, an FDA-approved first-line drug for the treatment of type 2 diabetes, has known beneficial effects on glucose metabolism
	        Intervention: Drug—Metformin
	• Experimental: Placebo
	    Placebo
	        Intervention: Drug—Placebo

Estimated enrollment	16
Sex/gender	All sexes eligible for study
Ages	60 years and older (adult, older adult)
NCT number	*NCT02308228*
Status	Completed
Official title	Novel Actions of Metformin to Augment Resistance Training Adaptations in Older Adults
Study start date	January 14, 2015
Study end date	December 14, 2017 (final data collection date for primary outcome measure)
Study description	The purpose of this study is to determine whether a commonly prescribed drug, metformin, can enhance the benefits seen during resistance exercise such as increased muscle mass and strength
Condition	Ageing

Study arms	• Experimental: Metformin
	    Participants will be randomized to receive metformin (1700 mg/day) for a period of 16 weeks and 2 weeks of metformin only followed by 14 weeks of continued metformin use in combination with progressive resistance training
	        Interventions:
	        • Behavioral: Progressive resistance training
	        • Drug: Metformin
	• Placebo comparator: Placebo, sugar pill
	    Participants will be randomized to receive placebo sugar pills (1700 mg/day) for a period of 16 weeks and 2 weeks of placebo only followed by 14 weeks of continued placebo use in combination with progressive resistance training. Placebos will be almost identical to the metformin medication
	        Intervention: Behavioral—Progressive resistance training

Estimated enrollment	109
Sex/gender	All sexes eligible for study
Ages	65 years and older (older adult)
NCT number	*NCT04264897*
Status	Recruiting
Official title	Does Insulin Sensitivity Impact the Potential of Metformin to Slow Ageing
Study start date	July 29, 2020
Study end date	April 30, 2024 (final data collection date for primary outcome measure)

Study characteristic	Description
Study description	Ageing is the number one risk factor for the majority of chronic diseases. There are no pharmaceutical treatments to slow ageing and prolong healthspan. The antidiabetic drug metformin is considered a likely pharmaceutical candidate to slow ageing. In this study, the investigators hypothesize that metformin treatment in subjects free of type 2 diabetes will improve insulin sensitivity and glucoregulation in insulin-resistant individuals, but will decrease insulin sensitivity and glucoregulation in insulin-sensitive subjects. Further, the investigators hypothesize that long-term metformin treatment will remodel mitochondria in a way that decreases mitochondrial function in subjects that are insulin sensitive, but improves mitochondrial function in subjects that are insulin resistant. The investigators will use a dual-site, 12-week drug intervention trial performed in a double-blind, placebo-controlled manner on 148 subjects recruited from two separate sites (Oklahoma Medical Research Foundation (OMRF) and University of Wisconsin-Madison (UWM)). After consent and initial subject screening for chronic disease, subjects will be stratified to insulin-sensitive (IS) or insulin-resistant (IR) groups. Over a 12-week intervention, half of each group will take metformin and half will take a placebo. Pre- and post-intervention, subjects will complete a series of procedures to assess insulin sensitivity, glucose regulation, and biomarkers of ageing. The same subjects will provide a skeletal muscle biopsy pre- and post-intervention to assess the change in mitochondrial function and mitochondrial remodeling with and without metformin treatment. By completion of this project, the investigators expect to provide evidence that helps further delineate who may benefit from metformin treatment to slow ageing
Condition	• Ageing • Insulin sensitivity • Chronic disease • Mitochondria • Insulin resistance

Study arms	• Experimental: Metformin
	    The investigators use a "ramp-up" dosing protocol in which the amount of metformin (Hunter Pharmacy) will begin at 500 mg/day in week 1, increase to 1000 mg/day in week 2, and then to 1500 mg/day in week 3, as tolerated. At week 3 and for the remaining 9 weeks, the dose will remain at 1500 mg/day, which is a standard clinical dose (1500–2000 mg/day). If a subject has gastrointestinal discomfort with a dose of 1500 mg/day, the investigators will lower the dose to 1000 mg/day. The investigators will split the dose with 1/2 given in the AM and 1/2 in the PM and taken with meals to minimize GI discomfort
	    Intervention: Drug—Metformin
	• Placebo comparator: Placebo
	    Subjects assigned to the placebo group will receive visually identical pills (silicified microcrystalline cellulose, Micosolle®, K30 povidone, sodium starch glycolate, and magnesium stearate)

Study characteristic	Description
	The same dosing schedule will be followed as for metformin. The investigators use a "ramp-up" dosing protocol in which the amount of placebo (Hunter Pharmacy) will begin at 500 mg/day in week 1, increase to 1000 mg/day in week 2, and then to 1500 mg/day in week 3, as tolerated. At week 3 and for the remaining 9 weeks, the dose will remain at 1500 mg/day. If a subject has gastrointestinal discomfort with a dose of 1500 mg/day, the investigators will lower the dose to 1000 mg/day. The investigators will split the dose with 1/2 given in the AM and 1/2 in the PM and taken with meals to minimize GI discomfort Intervention: Drug—Placebo oral tablet
Estimated enrollment	148
Sex/gender	All sexes eligible for study
Ages	40–75 years (adult, older adult)
NCT number	*NCT03072485*
Status	Completed
Official title	Phase 1 Study of the Effects of Combining Topical FDA-Approved Drugs on Age-Related Pathways on the Skin of Healthy Volunteers
Study start date	March 1, 2017
Study end date	February 22, 2019 (final data collection date for primary outcome measure)
Study description	The primary endpoint of the study is the profile of differences in transcript levels of age-associated genes such as those in the lamin-A, insulin-like growth factor (IGF), and NFKB pathways as well as noncoding RNAs in topical agent-exposed arm skin versus placebo-exposed arm skin in healthy volunteers. The secondary endpoints include (1) differences in skin wrinkling using a 4-point Likert scale for wrinkle severity between placebo and topical agent-exposed arm skin after 4 weeks of usage and (2) the type and severity of adverse events, both systemic and skin localized after exposure to both topical agent and placebo vehicle cream
Condition	Ageing

Study arms	• Sirolimus, metformin, and diclofenac
	   First five enrolled participants
	     Interventions:
	     • Drug: Sirolimus
	     • Drug: Metformin
	     • Drug: Diclofenac
	• Metformin and diclofenac
	   Sixth to tenth enrolled participants
	     Interventions:
	     • Drug: Metformin
	     • Drug: Diclofenac

Estimated enrollment	10

Study characteristic	Description
Sex/gender	Sexes eligible for study: Female
Ages	55 years and older (adult, older adult)
NCT number	*NCT03996538*
Status	Completed
Official title	Vaccination Efficacy with Metformin in Older Adults: A Pilot Study
Study start date	June 5, 2019
Study end date	February 4, 2020 (final data collection date for primary outcome measure)
Study description	With ageing the immune system gets weaker. This makes older adults more susceptible to influenza (flu). Vaccinations help to prevent infection from the flu virus; however, the immune system of older adults do not respond as well to vaccines compared to young adults and thus are not as well protected from the complications from the flu. This research is being done to determine if metformin, an FDA-approved diabetes medication, is effective at enhancing immune responses to flu vaccine in older men and women. Participants will be randomly assigned to either metformin or placebo treatment for a total of 22 weeks. Participants will be vaccinated with high-dose flu vaccine after 12 weeks of treatment. Immune responses will be evaluated throughout the study at six time points
Condition	• Ageing • Age-related immunodeficiency • Vaccine response impaired

Study arms	• Experimental: Metformin hydrochloride extended-release tablets
	    Patients will consume three tablets of 500 mg metformin hydrochloride extended-release tablets daily (1500 mg/day (after 3-week dose gradation))
	        Interventions:
	        • Drug: Metformin hydrochloride extended-release tablets
	        • Biological: Influenza vaccine
	• Placebo comparator: Placebo
	    Patients will consume three identical placebo tablets (after similar 3-week dose gradation)
	        Intervention: Biological—Influenza vaccine

Estimated enrollment	26
Sex/gender	All sexes eligible for study
Ages	65 years and older (older adult)
NCT number	*NCT01765946*
Status	Completed
Official title	Effects of Metformin on Longevity Gene Expression and Inflammation and Prediabetic Individuals. A Placebo-Controlled Trial
Study start date	June 2010
Study end date	March 2013 (final data collection date for primary outcome measure)

Study characteristic	Description
Study description	Prediabetes, a condition characterized by hyperglycemia, is associated with increased cardiovascular risk and reduced life expectancy, as compared to the general population. AMP-activated protein kinase (AMPK) is an enzyme that plays a key role in cellular energy homeostasis and metabolism, and recently it has been demonstrated that AMPK regulates ageing pathways as well. AMPK is susceptible to modulation through pharmacologic (e.g., metformin) and non-pharmacologic (e.g., physical exercise) interventions. This clinical trial aims to describe the effects of the AMPK pathway on longevity genes and inflammation in the setting of prediabetes in vivo and in vitro. To this end, the investigators will compare treatment with metformin (500 mg tid) for 2 months versus placebo in prediabetic subjects. The investigators will assess expression of longevity genes SIRT1, p66Shc, p53, and mTOR in peripheral blood mononuclear cells (PBMCs) ex vivo. The investigators will evaluate monocyte polarization by flow cytometry, according to the expression of surface antigens (CD68, CCR2, CD163, CD206, CX3CR1) to determine the prevalence of pro- or anti-inflammatory cells. Inflammatory cytokines (TNF-alpha, MCP-1, IL-1, IL-6, IL-10, CCL12) will also be determined. In the in vitro study, the investigators will evaluate the effects of AMPK activation or inhibition on longevity gene and protein expression
Condition	• Insulin resistance • Prediabetes • Ageing • Inflammation

Study arms	• Experimental: Metformin
	    Metformin tablets 500 mg tid for 2 months
	      Intervention: Drug—Metformin
	• Placebo comparator: Placebo
	    Placebo tables tid for 2 months
	      Intervention: Drug—Placebo

Estimated enrollment	38
Sex/gender	All sexes eligible for study
Ages	40–75 years (adult, older adult)
NCT number	NCT04221750
Status	Recruiting
Official title	Lifestyle Intervention Plus Metformin to Treat Frailty in Older Veterans with Obesity
Study start date	May 14, 2021
Study end date	September 30, 2025 (final data collection date for primary outcome measure)
Study description	The continuing increase in prevalence of obesity in older adults including many older Veterans has become a major health concern. The clinical trial will test the central hypothesis that a multicomponent intervention consisting of lifestyle therapy (diet-induced weight loss and exercise training) plus metformin will be the most effective strategy for reversing sarcopenic obesity and frailty in older Veterans with obesity
Condition	• Frailty • Sarcopenic obesity • Ageing

Study characteristic	Description

Study arms	• Experimental: Lifestyle therapy plus metformin
	   Diet-induced weight loss and exercise training plus metformin 1 gm bid
	     Interventions:
	     • Behavioral: Lifestyle therapy
	     • Drug: Metformin hydrochloride
	• Placebo comparator: Lifestyle therapy plus placebo
	   Diet-induced weight loss and exercise training plus placebo
	     Interventions:
	     • Behavioral: Lifestyle therapy
	     • Drug: Placebo
	• Active comparator: Healthy lifestyle plus metformin
	   Healthy lifestyle and metformin 1 gm bid
	     Interventions:
	     • Drug: Metformin hydrochloride
	     • Behavioral: Healthy lifestyle

Estimated enrollment	114
Sex/gender	All sexes eligible for study
Ages	65–85 years (older adult)
NCT number	NCT02570672
Status	Active, not recruiting
Official title	Metformin for Preventing Frailty in High-Risk Older Adults
Study start date	April 2016
Study end date	October 2023 (final data collection date for primary outcome measure)
Study description	Frailty is a geriatric syndrome that leads to poor health outcomes in older adults, such as falls, disability, hospitalization, institutionalization, and death. Due to the dramatic growth in the US ageing population and the healthcare costs associated with frailty (estimated at more than $18 billion per year), frailty is a major healthcare problem. There has been little research into potential pharmacologic interventions that would delay or reduce the incidence of frailty. Thus, the major goal of this study is to test metformin as a novel intervention for the prevention of frailty. The investigators propose that diabetes/insulin resistance and inflammation are major contributors to frailty and that the use of metformin to modulate diabetes/insulin resistance and inflammation will prevent and/or ameliorate the progression of frailty
Condition	Frailty

Study arms	• Experimental: Metformin
	   Subjects will be randomized to metformin (titrated up to 1000 mg twice daily, as tolerated)
	     Intervention: Drug—Metformin
	• Placebo comparator: Placebo
	   Subjects will be randomized to metformin (titrated up to 1000 mg twice daily, as tolerated) versus placebo
	     Intervention: Drug—Placebo

Estimated enrollment	125
Sex/gender	All sexes eligible for study
Ages	65–90 years (older adult)

Appendix 4: Characteristics of Studies Examining the Effects of Rapamycin and Its Derivatives on Healthy Individuals and Individuals with Age-Related Diseases

	Study location	Study design	Intervention	Comparator	Study duration, days	Cohort size, N	Age, years[a]	Study participants by sex, N	Health status
Studies on healthy individuals									
Boni et al. (2012)	USA	Crossover	Temsirolimus	Placebo + moxifloxacin	2	58	18.0–50.0	0 women, 58 men	Healthy
Chung et al. (2019)	USA	RCT	Rapamycin	Placebo	240	36	>40.0	28 women, 8 men	No major morbidities but evidence of age-related photoageing and dermal volume loss
Dickinson et al. (2013)	USA	Crossover	Rapamycin	Self	2	6	26.0	3 women, 3 men	Healthy
Drummond et al. (2009)	USA	RCT	Rapamycin	No intervention	1	Overall 15, intervention 7, control 8	29.0	0 women, 15 men	Healthy
Gundermann et al. (2014)	USA	RCT	Rapamycin + blood flow restriction	Blood flow restriction exercise	2	16	25.5	0 women, 16 men	Healthy
Hörbelt et al. (2020)	Germany	Pilot study	Everolimus	Self	15	Overall 19, intervention one 6, intervention two 7, intervention three 6	Overall 28.2, intervention one 27.8, intervention two 27.7, intervention three 28.8	Intervention one: 0 women, 6 men; intervention two: 0 women, 7 men; intervention three: 0 women, 6 men	Healthy

	Study location	Study design	Intervention	Comparator	Study duration, days	Cohort size, N	Age, years[a]	Study participants by sex, N	Health status
Krebs et al. (2007)	Austria	Crossover study	Rapamycin	Placebo	2	Overall 11, intervention one 8, intervention two 3	Overall 28.3, intervention one 29.0, intervention two 26.0	Intervention one: 0 women, 8 men; intervention two: 0 women, 3 men	Healthy, without a family history of diabetes, dyslipidemia, or conditions related to insulin resistance, and not taking any medication
Kraig et al. (2018)	USA	RCT phase 1	Rapamycin	Placebo	120	Overall 8, intervention 4, control 4	81.0–95.0	Intervention: 0 women, 4 men; control: 0 women, 4 men	All chronic diseases (e.g., hypertension and coronary artery disease) clinically stable
Kraig et al. (2018)	USA	RCT phase 2	Rapamycin	Placebo	60	Overall 17, intervention 7, control 10	70.0–95.0; intervention group 74.8[b]	Intervention: 2 women, 5 men; control: 5 women, 5 men	All chronic diseases (e.g., hypertension and coronary artery disease) clinically stable
Mannick et al. (2021)	New Zealand	RCT phase 2b (part 1)	RTB101	Placebo	..	Overall 179, intervention one 61, intervention two 58, control 60	Overall 74.9, intervention one 74.0, intervention two 76.5, control 74.4	Intervention one: 28 women, 33 men; intervention two: 27 women, 31 men; control: 24 women, 36 men	With asthma, type 2 diabetes, chronic obstructive pulmonary disease, or congestive heart failure

Study	Country	Study type	Drug	Control	N	Sample size	Age	Sex	Conditions
Mannick et al. (2021)	USA	RCT phase 2b (part 2)	RTB101	Control one, RTB101 + everolimus; control two, placebo	168	Overall 473, intervention one 118, intervention two 120, control one 115, control two 120	Overall 73.3, intervention one 73.1, intervention two 73.0, control one 73.9, control two 73.2	Intervention one: 66 women, 52 men; intervention two: 59 women; 61 men; control one: 58 women, 57 men; control two: 67 women, 53 men	With asthma, type 2 diabetes, chronic obstructive pulmonary disease, or congestive heart failure
Mannick et al. (2021)	New Zealand and Australia	RCT phase 3	RTB101	Placebo	80	Overall 1021, intervention 511, control 510	Overall 72.8, intervention 72.6, control 73.1	Intervention: 292 women, 219 men; control: 286 women, 224 men	Without chronic obstructive pulmonary disease or other significant pulmonary disease other than asthma, current evidence of unstable cardiac condition or other serious or unstable medical disorder
Mannick et al. (2014)	New Zealand and Australia	RCT	RAD001	Placebo	70	Overall 218, intervention one 53, intervention two 53, intervention three 53, control 59	Overall 71.3, intervention one 70.8, intervention two 72.0, intervention three 71.4, control 71.1	Intervention one: 19 women, 34 men; intervention two: 26 women, 27 men; intervention three: 21 women, 32 men; control: 28 women, 31 men	Without unstable medical conditions
Studies on individuals with age-related diseases									
Bruyn et al. (2008)	Europe and USA	RCT	Everolimus	Placebo	84	Overall 121, intervention 61, control 60	Overall 54.5[b]	Overall: 93 women, 28 men[b]	Rheumatoid arthritis

	Study location	Study design	Intervention	Comparator	Study duration, days	Cohort size, N	Age, years[a]	Study participants by sex, N	Health status
Dugel et al. (2012)	USA	RCT	Rapamycin (subconjunctival)	Rapamycin (intravitreal)	365	Overall 50, subconjunctival 25, intravitreal 25	Overall 63.5, subconjunctival 64.0, intravitreal 63.0	Subconjunctival: 13 women, 12 men; intravitreal: 14 women, 11 men	Diabetic macular oedema
Gensler et al. (2018)	USA	RCT	Rapamycin	Lidocaine	365–730	Overall 52, intervention 27, control 25	Overall 79.1, intervention 78.5, control 79.8	Intervention: 17 women, 10 men; control: 12 women, 13 men	Geographic atrophy in age-related macular degeneration
Minturn et al. (2021)	USA	RCT	Rapamycin	Bevacizumab or aflibercept	182.5	Overall 40, intervention 20, control 20	Overall 81.8, intervention 84.5, control 79.2	..	Persistent exudative age-related macular degeneration
Nussenblatt et al. (2010)	USA	RCT phase 1/2	Rapamycin	Observation	182.5	Overall 13, intervention (rapamycin) 3, intervention (daclizumab) 4, intervention (infliximab) 3, control (observation) 3	Overall 80.3, intervention (rapamycin) 87.0, intervention (daclizumab) 79.2, intervention (infliximab) 77.3, control (observation) 78.0	Intervention (rapamycin): 3 women, 0 men; intervention (daclizumab): 4 women, 0 men; Intervention (infliximab): 2 women, 1 man; control: 3 women, 0 men	Age-related macular degeneration with choroidal neovascularization
Palma et al. (2022)	USA	RCT	Rapamycin	Placebo	406	Overall 47, intervention 35, control 12	Overall 58.5 intervention 59.0, control 58.0	Intervention: 14 women, 21 men; control: 6 women, 6 men	Parkinsonian-predominant or cerebellar predominant multiple system atrophy
Petrou et al. (2014)	USA	RCT	Rapamycin	Fellow eye, no treatment	730	6	74.3	2 women, 44 men	Geographic atrophy

Seyfarth et al. (2013)	Germany	RCT	Everolimus[c] conventional treatment	None	180	10	51.6	6 women, 4 men	Pulmonary hypertension
Wen et al. (2019)	China	RCT	Rapamycin + conventional treatment	Methotrexate, leflunomide, hydroxychloroquine, or thalidomide	168	Overall 62, intervention 42, control 20	Intervention 50.3, control 51.8	Intervention: 34 women, 8 men; control: 17 women, 3 men	Rheumatoid arthritis
Bruyn et al. (2008)	Europe and USA	RCT	Everolimus	Placebo	84	Overall 121, intervention 61, control 60	Overall 54.5[b]	Overall: 93 women, 28 men[b]	Rheumatoid arthritis

Source: Lee et al. (2024)

Notes: *RCT* = randomized controlled trial

[a]Ages are mean values for all studies unless otherwise indicated, except for the study of Palma and colleages (2022) which shows median values

[b]Exact data unavailable for the overall and control groups

[c]Data were not provided separately for control and intervention groups

Appendix 5: Clinical Trials of Senotherapeutics Registered on https://www.clinicaltrials.gov as of February 18, 2024

Study characteristic	Description
	Description
NCT number	NCT05276895
Status	Not yet recruiting
Official title	Evaluation of the Efficacy of Natural Senolytic Agents and NLRP3 Inhibitors in Treatment of Osteoarthritis: Randomized, Double Blinded, Placebo Controlled Trial
Study start date	March 1, 2023
Study end date	September 1, 2023 (final data collection date for primary outcome measure)
Study description	Osteoarthritis is a progressive degenerative disease of the joint leading to cartilage damage, pain, and loss of function affecting an estimated 250 million people worldwide and 27 million people in the USA. Currently, there are no effective FDA-approved therapies that are disease-modifying interventions to block the joint destruction pathway because of osteoarthritis. The most prevalent first-line treatment for OA is to mitigate pain and restore function with a combination of weight management, physical therapy, mind-body exercises, and analgesia with paracetamol or NSAIDs (topical or oral). Another prominent treatment strategy is the use of intra-articular corticosteroids (CS) to reduce pain and inflammation via targeting production of interleukins, leukotrienes, and prostaglandins However, the palliative effects of CS for OA are often short term and can potentially lead to chondral fissuring and promotion of dose-independent structural changes in cartilage, and there are no consistent reports of efficacy One novel potential and appealing approach for treating osteoarthritis is through the local and systemic elimination of senescent cells. Senescent cell burden increases significantly with age and has been shown to promote several age-related pathologies including degenerative joint conditions. Senescent cells are non-proliferative, resistant to apoptosis, and secrete pro-inflammatory factors that promote disease and systemic ageing. Cellular senescence can be induced by a variety of extrinsic and intrinsic signals that leads to the production of a collection of various pro-inflammatory cytokines and other factors that initiate senescence in neighboring cells and promote disease and tissue dysfunction. Thus, senescent cells and their senescence-associated secretory phenotype profiles likely play a role in both the clinical manifestation of OA (pain) and disease pathogenesis (tissue dysfunction and cartilage degradation The overall safety and efficacy of several senolytic drugs to treat chronic diseases have been demonstrated in several preclinical studies and more recently in phase 1–2 clinical trials for OA. However, there are no encouraging results with the use of natural senescent agents such as quercetin or fisetin as disease-modifying agents in OA; therefore, our study will investigate the effect of a combination of natural senescent agents and NLRP3 inhibitors on inflammation

Study characteristic	Description
	Inflammasomes play a crucial role in innate immunity by serving as signaling platforms that deal with a plethora of pathogenic products and cellular products associated with stress and damage. By far, the best studied and most characterized inflammasome is NLRP3 inflammasome, which consists of NLRP3 (nucleotide-binding domain leucine-rich repeat (NLR) and pyrin domain containing receptor 3). The hyperactivation of NLRP3 inflammasome is involved in a wide range of inflammatory diseases. The search and development of anti-inflammatory drugs from natural sources of plants has received extensive attention. Licorice extract has high activity and wide therapeutic effects. It has been reported that glycyrrhizin could ameliorate fibrosis and inflammation via inhibiting NLRP3 inflammasome activation and NF-κB signaling pathway Our aim is to determine the efficacy of natural senolytic agents and NLRP3 inflammasome inhibitors for reducing knee symptoms and effusion-synovitis in patients with symptomatic knee osteoarthritis and knee effusion-synovitis
Condition	Osteoarthritis

Study arms	• Active comparator: quercetin + fisetin
	    20 participants with symptomatic knee osteoarthritis and ultrasonography defined effusion-synovitis will take natural senolytic agents
	    The senolytic agents act by hit-and-run strategy; therefore, intermittent dosing regimens will be applied. 1250 mg/day quercetin + 1000 mg/day fisetin for three consecutive days every 3 weeks, over 12 weeks
	        Intervention: Dietary supplement—Quercetin and fisetin tablet
	• Active comparator: Quercetin + fisetin + glycyrrhizin
	    20 participants with symptomatic knee osteoarthritis and ultrasonography-defined effusion-synovitis will take quercetin 1250 mg + fisetin 1000 mg for three consecutive days followed by 100 mg/day glycyrrhizin for 1 week every 3 weeks over 12 weeks
	        Intervention: Dietary supplement—Quercetin capsule/tablet, fisetin capsule/tablet and glycyrrhizin capsules
	• Placebo comparator: Placebo
	    Placebo-controlled group
	        Intervention: Other—Placebo

Estimated enrollment	60
Sex/gender	All sexes eligible for study
Ages	40–80 years (adult, older adult)
NCT number	NCT05025956
Status	Recruiting
Official title	The Use of Senolytic Agent to Improve the Benefit of Platelet-Rich Plasma and Losartan for Treatment of Femoroacetabular Impingement and Labral Tear: A Pilot Study
Study start date	October 24, 2021
Study end date	August 2023 (final data collection date for primary outcome measure)

Study characteristic	Description
Study description	The purpose is to explore the possible benefit of administration of fisetin (a senolytic agent) to improve the benefit of platelet-rich plasma and losartan for treatment of femoroacetabular impingement and labral tear We believe that giving fisetin, a senolytic agent, will improve the benefit of PRP by eliminating senescent cells and senescence-associated secretory phenotype (SASP), known to exist in PRP. The main objectives of this study are to determine if pre- and postoperative administration of a senolytic agent will improve the beneficial effects of PRP when used in conjunction with surgical treatment of FAI and/or labral tear, to determine whether pre- and postoperative administration of fisetin is associated with adverse events and to determine if pre- and postoperative administration of fisetin leads to a decrease in systemic senescence, serum SASP, and fibrotic markers Patients suffering from femoroacetabular impingement and labral tear, who are planning to undergo hip arthroscopy combined with standard of care intraoperative PRP injection and postoperative losartan administration, will be recruited from the clinical practice of the principal clinical investigator or his designee at The Steadman Clinic (TSC)
Condition	Femoroacetabular Impingement

Study arms	• Experimental: Fisetin group (investigational group)
	   20 mg/kg of fisetin per day for days 1 and 2 prior to surgery and days 33, 34, 63, 64, 93, and 94 post-surgery
	   (The pills are 100 mg each. For example, if a participant weighs 160 pounds (about 73 kg), the participant will need to take 15 pills per day)
	      Intervention: Drug—Fisetin
	• Placebo comparator: Placebo group (control group)
	   20 mg/kg of placebo per day for days 1 and 2 prior to surgery and days 33, 34, 63, 64, 93, and 94 post-surgery
	   (The pills are 100 mg each. For example, if a participant weighs 160 pounds (about 73 kg), the participant will need to take 15 pills per day)
	      Intervention: Drug—Placebo

Estimated enrollment	100
Sex/gender	All sexes eligible for study
Ages	18–80 years (adult, older adult)
NCT number	NCT04815902
Status	Active, not recruiting
Official title	The Use of Senolytic and Anti-fibrotic Agents to Improve the Beneficial Effect of Bone Marrow Stem Cells for Osteoarthritis
Study start date	May 18, 2021
Study end date	February 1, 2025 (final data collection date for primary outcome measure)

Study characteristic	Description
Study description	This is a prospective, randomized, double-blind, active control clinical trial to evaluate the safety and efficacy of a senolytic agent (fisetin) and an anti-fibrotic agent (losartan), used independently and in combination, to improve beneficial effect demonstrated by the active control which is to be injection of autologous bone marrow aspirate concentrate (BMAC) into an osteoarthritic knee This is a prospective, randomized, double-blind, active control clinical trial to evaluate the safety and efficacy of a senolytic agent (fisetin) and an anti-fibrotic agent (losartan), used independently and in combination, to improve beneficial effect demonstrated by the active control which is to be injection of autologous bone marrow aspirate concentrate (BMAC) into an osteoarthritic knee. One hundred subjects with symptomatic unilateral or bilateral knee osteoarthritis (Kellgren-Lawrence grades II–IV) will be randomized into one of four arms (1:1:1:1). All subjects will receive an injection of BMAC. Group $1 - n = 25$, control (BMA concentrate + fisetin placebo + losartan placebo); group $2 - n = 25$, BMA concentrate + fisetin placebo + active losartan; group $3 - n = 25$, BMA concentrate + active fisetin + losartan placebo; group $4 - n = 25$, BMA concentrate + active fisetin + active losartan
Condition	Osteoarthritis, knee

Study arms	• Experimental: Active fisetin and active losartan
	    Losartan 12.5 mg, PO, BID beginning the first day after BMA concentrate injection and continuing for 30 days. Fisetin 20 mg/kg taken a total of 4 days prior to BMA concentrate injection (-32 and -31 and -3 and -2) and then again after BMA concentrate injection for a total of 6 days (32 and 33, 61 and 62, and 90 and 91)
	    Interventions:
	    • Drug: Fisetin
	    • Drug: Losartan
	• Active comparator: Active fisetin and losartan placebo
	    Losartan placebo 12.5 mg, PO, BID beginning the first day after BMA concentrate injection and continuing for 30 days. Fisetin 20 mg/kg taken a total of 4 days prior to BMA concentrate injection (-32 and -31 and -3 and -2) and then again after BMA concentrate injection a for a total of 6 days (32 and 33, 61 and 62, and 90 and 91)
	    Interventions:
	    • Drug: Fisetin
	    • Drug: Placebo losartan
	• Active comparator: Fisetin placebo and active losartan
	    Losartan 12.5 mg, PO, BID beginning the first day after BMA concentrate injection and continuing for 30 days. Fisetin placebo 20 mg/kg taken a total of 4 days prior to BMA concentrate injection (-32 and -31 and -3 and -2) and then again after BMA concentrate injection for a total of 6 days (32 and 33, 61 and 62, and 90 and 91)
	    Interventions:
	    • Drug: Losartan
	    • Drug: Placebo fisetin
	• Placebo comparator: Control
	    Losartan placebo 12.5 mg, PO, BID beginning the first day after BMA concentrate injection and continuing for 30 days. Fisetin placebo 20 mg/kg taken a total of 4 days prior to BMA concentrate injection (-32 and -31 and -3 and -2) and then again after BMA concentrate injection a for a total of 6 days (32 and 33, 61 and 62, and 90 and 91)
	    Interventions:
	    • Drug: Placebo losartan
	    • Drug: Placebo fisetin

Study characteristic	Description
Estimated enrollment	100
Sex/gender	All sexes eligible for study
Ages	40–85 years (adult, older adult)
NCT number	NCT04063124
Status	Completed
Official title	Pilot Study to Investigate the Safety and Feasibility of Senolytic Therapy to Modulate Progression of Alzheimer's Disease (SToMP-AD)
Study start date	February 14, 2020
Actual study completion date	January 30, 2023
Study description	The purpose of this pilot study is to evaluate whether a combination of two drugs, dasatinib (D) and quercetin (Q) [D + Q], penetrate the brain using cerebrospinal fluid (CSF) in older adults with early Alzheimer's disease (AD). This combination of drug therapy has been shown to affect dying cells in humans with other chronic illnesses and in Alzheimer's mice models. The study team want to know if this combination of medications will reach the brain in order to evaluate if this intervention may be effective for treating AD symptoms in future studies. This is also known as a "proof-of-concept" study
Condition	Alzheimer's disease
Study arms	Experimental: Intermittent D + Q Senolytic treatment in five individuals with early AD to determine levels of drug that reach the central nervous system (CNS) by collecting cerebral spinal fluid (CSF), and begin collecting initial data on target engagement of senescent cells, AD-related markers, and AD-relevant outcomes for future trials Intervention: Drug—Dasatinib + quercetin
Actual enrollment	5
Sex/gender	All sexes eligible for study
Ages	65 years and older (older adult)
NCT number	NCT04210986
Status	Completed
Official title	Senolytic Drugs Attenuate Osteoarthritis-Related Articular Cartilage Degeneration: A Clinical Trial
Actual study start date	January 6, 2020
Actual study end date	June 1, 2022 (final data collection date for primary outcome measure)

Study characteristic	Description
Study description	This is a phase 1/phase 2 randomized, double-blind, placebo-controlled clinical trial that will be conducted at The Steadman Clinic (TSC) and Steadman Philippon Research Institute (SPRI). The purpose of this study is to evaluate the clinical efficacy of fisetin (FIS), a dietary supplement, in symptomatic knee osteoarthritis (OA) patients. Key aspects of this proposal include the investigator's well-developed methodologies to measure and compare systemic senescence-associated secretory phenotype (SASP) including inflammatory biomarkers and senescent cells and collect magnetic resonance images, self-reported outcomes, physical performance, and other objective clinical data. Given the drug FIS has been empirically demonstrated to reduce senescent cell burden, the main objective(s) are to determine (1) the safety of FIS during dosing and (2) whether FIS reduces senescent cells, pro-inflammatory and cartilage degenerating SASP markers and reduces OA symptoms leading to improved joint health and function
Condition	Osteoarthritis, knee

Study arms	• Experimental: Fisetin
	    Fisetin 100 mg capsules (~20 mg/kg/day) will be administered orally for two consecutive days (days 1 and 2) followed by 28 days off. A second course will be given for two consecutive days (days 31 and 32)
	        Intervention: Dietary supplement—Fisetin
	• Placebo comparator: Placebo
	    Placebo capsules will be administered orally for two consecutive days (days 1 and 2) followed by 28 days off. A second course will be given for two consecutive days (days 31 and 32)
	        Intervention: Drug—Placebo oral capsule

Actual enrollment	75
Sex/gender	All sexes eligible for study
Ages	40–80 years (adult, older adult)
NCT number	NCT04685590
Status	Recruiting
Official title	Phase II Clinical Trial to Evaluate the Safety and Feasibility of Senolytic Therapy in Alzheimer's Disease
Study start date	December 22, 2021
Study end date	January 2027 (final data collection date for primary outcome measure)
Study description	This study is a phase 2 multi-site, randomized, double-blind placebo-controlled trial to determine safety, feasibility, and efficacy of senolytics in older adults with amnestic mild cognitive impairment (MCI) or early-stage AD (Clinical Dementia Rating (CDR) = 0.5 or 1) who are tau PET positive
Condition	• Alzheimer's disease, early onset • Mild cognitive impairment

Study characteristic	Description

Study arms	• Experimental: Treatment
	    Dasatinib (D) is given as (1) 100 mg capsule daily for two consecutive days (Sprycel®, Bristol Myers Squibb). Quercetin (Q) will be given as (4) 250 mg capsules daily (total 1000 mg daily) for the same two consecutive days (Thorne Research). Both are administered orally
	      Intervention: Drug—Dasatinib + quercetin
	• Placebo comparator: Placebo
	    Matching placebo capsules following the same administration protocol as the experimental treatment—administered once daily (first dose of each cycle will be given, supervised, at the clinic visit; the second dose will be taken at home) for two consecutive days followed by a 13-day (±2 day) no-drug period for 12 consecutive weeks for six rounds of administration
	      Intervention: Other—Placebo capsules

Estimated enrollment	48
Sex/gender	All sexes eligible for study
Ages	65 years and older (older adult)
NCT number	NCT04733534
Status	Recruiting
Official title	SEN-SURVIVORS: An Open-Label Intervention Trial to Reduce Senescence and Improve Frailty in Adult Survivors of Childhood Cancer
Study start date	June 6, 2022
Study end date	July 2023 (final data collection date for primary outcome measure)
Study description	This is a first-in survivor pilot study with the goal of establishing preliminary evidence of efficacy, safety, and tolerability of two senolytic regimens to reduce markers of cellular senescence (primary outcome: p16^INK4a) and improve frailty (primary outcome: walking speed) in adult survivors of childhood cancer. If successful, this pilot would provide the preliminary evidence needed for a phase 2, randomized, placebo-controlled trial to establish efficacy *Primary objective* • The primary aim of this proposal is to test the efficacy of two, short-duration senolytic regimens: (1) combination of dasatinib plus quercetin and (2) fisetin alone, to improve walking speed and decrease senescent cell abundance in blood (p16^INKA) • Primary endpoints of this trial will be change in walking speed and senescent cell abundance in blood (p16^INK4A) determined at baseline and again at 60 days, within an individual arm. Extended follow-up at 150 days will assess the permanence of change after completion of the trial. Secondary endpoints of this trial will be effect of intervention on additional measures of frailty (beyond walking speed; Fried criteria) and on other cell senescence markers, markers of inflammation, insulin resistance, bone resorption, and cognitive function *Secondary objectives* The secondary aim is to test the safety and tolerability of two different senolytic therapies Exploratory objectives • To compare the efficacy of the two senolytic regimens in improving walking speed and decreasing senescent cell abundance • To evaluate the longitudinal pattern in measures of frailty

Study characteristic	Description
Condition	• Frailty • Childhood cancer

Study arms	• Active comparator: Dasatinib plus quercetin
	    Day 0 (30 per arm, randomization stratified by sex and age)
	    At the visit on day 7, blood CD3+ T lymphocyte p16^INK4A mRNA and other markers of inflammation and senescence will be accessed to verify that senescent cells have been cleared by the intervention
	    Post-treatment follow-up will occur on days 60 for primary endpoints and day 150 for secondary evaluation. Day 150 will assess the permanence of change after completion of the trial
	        Intervention: Drug—Dasatinib plus Quercetin
	• Active comparator: Fisetin
	    Day 0 (30 per arm, randomization stratified by sex and age)
	    At the visit on day 7, blood CD3+ T lymphocyte p16INK4A mRNA and other markers of inflammation and senescence will be accessed to verify that senescent cells have been cleared by the intervention
	    Post-treatment follow-up will occur on days 60 for primary endpoints and day 150 for secondary evaluation. Day 150 will assess the permanence of change after completion of the trial
	        Intervention: Drug—Fisetin

Estimated enrollment	60
Sex/gender	All sexes eligible for study
Ages	18 years and older (adult, older adult)
NCT number	NCT02848131
Status	Enrolling by invitation
Official title	Senescence, Frailty, and Mesenchymal Stem Cell Functionality in Chronic Kidney Disease: Effect of Senolytic Agents
Study start date	July 2016
Study end date	April 2025 (final data collection date for primary outcome measure)
Study description	The proposed studies will examine cellular senescence and the effect of senolytic therapy on senescent cell burden, frailty, and adipose-derived mesenchymal stem cell function in individuals with diabetic chronic kidney disease
Condition	Chronic kidney disease

Study arms	• No intervention: Group 1—Observational
	    Observational only
	• Active comparator: Group 2—Dasatinib and quercetin
	    The drugs dasatinib and quercetin will be used in this arm
	        Interventions:
	      • Drug: Group 2—Dasatinib
	      • Drug: Group 2—Quercetin

Estimated enrollment	30
Sex/gender	All sexes eligible for study
Ages	40–80 years (adult, older adult)

Study characteristic	Description
NCT number	NCT04903132
Status	Recruiting
Official title	Cellular Senescence and Its Contribution to COVID-19 Long-Hauler Syndrome
Study start date	March 1, 2021
Study end date	May 2023 (final data collection date for primary outcome measure)
Study description	The purpose of this study is to test if senescent cells and their secretome contribute to long-hauler syndrome and if a clinical trial of senolytic drugs, which selectively eliminate senescent cells, should be initiated
Condition	SARS-CoV-2 infection
Study arms	Long-hauler cohort: Patients must be at least 18 years or older, have a positive PCR or antibody test within 6 months, and report any ongoing symptoms associated with post-COVID or long-hauler syndrome Control cohort: Patients must be at least 18 years or older and have not had a known case of COVID-19 or long-hauler syndrome COVID control cohort: Patients must be at least 18 years or older and have had COVID-19 but no known long-hauler syndrome
Estimated enrollment	200
Sex/gender	All sexes eligible for study
Ages	18 years and older (adult, older adult)
NCT number	NCT05505747
Status	Not yet recruiting
Official title	Enhancing Recovery Through a Combined Mechanobiologic Intervention Following Meniscus Repair
Study start date	January 2025
Study end date	February 2026 (final data collection date for primary outcome measure)
Study description	Arthroscopic meniscal procedures are the most commonly performed orthopedic procedure in the USA affecting 15% of Americans ages 10–65 years. Meniscus injury is also known to increase the risk of posttraumatic osteoarthritis (PTOA). The current randomized clinical trial will test a novel intervention after meniscal repair that combines an oral senolytic fisetin and real-time biofeedback program to restore joint loading and subsequent return to activity Arthroscopic meniscal procedures are the most commonly performed orthopedic procedure in the USA affecting 15% of Americans ages 10–65 years. Meniscus injury is also known to increase the risk of posttraumatic osteoarthritis (PTOA). The current randomized clinical trial will test a novel mechanobiologic intervention after meniscal repair that combines an oral senolytic fisetin and real-time biofeedback program to restore joint loading and subsequent return to activity. The premise for this program is derived from growing evidence that under-loading early after an orthopedic surgery is a major contributing factor to future PTOA development. Despite the need to assess how much force the patient can produce in various exercises to better inform progression and progress, clinicians still have few tools. By providing the real-time feedback during all exercises, the program will promote the recovery of muscle function as well which is critical for normal joint loading

Study characteristic	Description
	The study will (1) evaluate the feasibility and acceptability of the mechanobiologic intervention following arthroscopic repair of meniscus tears; (2) determine if the mechanobiologic intervention improves physical function, patient-reported outcomes, loading and muscle strength, and biomechanical symmetry in comparison to standard of care physical therapy and oral placebo; and (3) compare cartilage composition at 1 year after meniscus surgery between the mechanobiologic intervention and a control group-treated standard of care physical therapy and oral placebo
Condition	• Meniscus tear • Meniscus derangement • Meniscus lesion • Meniscus disorder

Study arms	• Experimental: Fisetin
	    Subjects assigned to the experimental group will take approximately 20 mg/kg/day of fisetin for two consecutive days, followed by a 28-day senescence washout period, and then another 2-day administration. Fisetin treatment begins at 8 weeks after surgery
	        Intervention: Drug—Fisetin
	• Placebo comparator: Placebo
	    Subjects assigned to the experimental group will take approximately 20 mg/kg/day of placebo (cornstarch) for two consecutive days, followed by a 28-day washout period, and then another 2-day administration. Placebo treatment begins at 8 weeks after surgery
	        Intervention: Drug—Placebo oral capsule

Estimated enrollment	38
Sex/gender	All sexes eligible for study
Ages	18–45 years (adult)
NCT number	NCT04313634
Status	Active, not recruiting
Official title	Targeting Cellular Senescence with Senolytics to Improve Skeletal Health in Older Humans: A Phase 2, Single-Center, 20-Week, Open-Label, Randomized Controlled Trial
Study start date	June 9, 2020
Study end date	June 2023 (final data collection date for primary outcome measure)
Study description	To determine if senolytic drugs reduce senescent cell burden and reduce bone resorption markers/increase bone formation markers in elderly women
Condition	Healthy

Study characteristic	Description

Study arms	• Experimental: Dasatinib plus quercetin treatment group
	    Subjects will receive dasatinib (D; 100 mg for 2 days) plus quercetin (Q; 1000 mg total daily for three consecutive days) taken orally on an intermittent schedule (starting every 28 days) with no-therapy periods in between dosing regimens, repeated every 28 days over 20 weeks, resulting in five total dosing periods throughout the entire intervention
	        Interventions:
	      • Drug: Dasatinib
	      • Drug: Quercetin
	• Experimental: Fisetin treatment group
	    Subjects will receive fisetin (F; ~20 mg/kg/day for three consecutive days) taken orally on an intermittent schedule (starting every 28 days) with no-therapy periods in between dosing regimens, repeated every 28 days over 20 weeks, resulting in five total dosing periods throughout the entire intervention
	        Intervention: Drug—Fisetin
	• No intervention: Untreated control group
	    Subjects will not receive any intervention

Estimated enrollment	120
Sex/gender	Female
Ages	60 years and older (adult, older adult)
NCT number	NCT05758246
Status	Not yet recruiting
Official title	Senolytics to Slow Progression of Sepsis (STOP-Sepsis) Trial
Study start date	April 1, 2023
Study end date	April 1, 2026 (final data collection date for primary outcome measure)
Study description	The long-term goal is to test the clinical efficacy of senolytic therapies to reduce progression to and severity of sepsis in older patients. The central hypothesis is that a threshold burden of SnCs predisposes to a SASP-mediated dysfunctional response to PAMPs, contributing to a disproportionate burden of sepsis in older patients. The study hypothesizes that timely treatment with fisetin will interrupt this pathway A multicenter, randomized, adaptive allocation clinical trial to identify the most efficacious dose of the senolytic fisetin to reduce the composite cardiovascular, respiratory, and renal sequential organ failure assessment score at 1 week and predict the probability of success of a definitive phase 3 clinical trial
Condition	• Sepsis • Acute infection • Organ failure

Study characteristic	Description

Study arms	• Experimental: Fisetin-dose 1
	    Elderly (≥65 years) patients admitted to the hospital with an acute infection and sequential organ failure assessment score sufficient for a diagnosis of sepsis will be given the first dose of fisetin
	        Intervention: Drug—Fisetin-dose 1
	• Experimental: Fisetin-dose 2
	    Elderly (≥65 years) patients admitted to the hospital with an acute infection and sequential organ failure assessment score sufficient for a diagnosis of sepsis will be given the second dose of fisetin
	        Intervention: Drug—Fisetin-dose 2
	• Placebo comparator: Placebo
	    Elderly (≥65 years) patients admitted to the hospital with an acute infection and sequential organ failure assessment score sufficient for a diagnosis of sepsis will receive placebo treatment
	        Intervention: Drug—Placebo

Estimated enrollment	220
Sex/gender	All sexes eligible for study
Ages	65 years and older (older adult)
NCT number	NCT04771611
Status	Active, not recruiting
Official title	COVFIS-HOME: A Phase 2 Placebo-Controlled Pilot Study in COVID-19 of Fisetin to Alleviate Dysfunction and Decrease Complications in At-Risk Outpatients
Study start date	July 14, 2021
Study end date	September 2023 (final data collection date for primary outcome measure)
Study description	The purpose of this study is to test whether fisetin, a senolytic drug, can assist in the reduction of complications in patients with COVID-19 infection To determine whether short-term treatment with fisetin reduces the rate of death and long-term complications related to COVID-19 and to determine the safety of treatment with fisetin in this patient population
Condition	• COVID-19 • Coronavirus infection

Study arms	• Experimental: Treatment group
	    Subjects will receive treatment drug (fisetin)
	        Intervention: Drug—Fisetin
	• Placebo comparator: Placebo
	    Subjects will receive placebo
	        Intervention: Drug—Fisetin

Estimated enrollment	150
Sex/gender	All sexes eligible for study
Ages	18 years and older (adult, older adult)
NCT number	NCT04537299
Status	Enrolling by invitation
Official title	COVID-FIS: A Phase 2 Placebo-Controlled Pilot Study in COVID-19 of Fisetin to Alleviate Dysfunction and Excessive Inflammatory Response in Older Adults in Nursing Homes

Study characteristic	Description
Study start date	January 25, 2022
Study end date	December 2023 (final data collection date for primary outcome measure)
Study description	The purpose of this study is to test whether fisetin, a senolytic drug, can assist in preventing an increase in the disease's progression and alleviate complications of coronavirus due to an excessive inflammatory reaction This study is a pilot, randomized, placebo-controlled, single-center study of fisetin in elderly nursing home participants with non-, mildly, or moderately symptomatic and confirmed SARS-CoV-2 infection
Condition	• COVID-19 • SARS-CoV infection

Study arms	• Experimental: Treatment group
	    Subjects will receive treatment drug (fisetin)
	      Intervention: Drug—Fisetin
	• Placebo comparator: Placebo group
	    Subjects will receive placebo
	      Intervention: Drug—Placebo

Estimated enrollment	150
Sex/gender	All sexes eligible for study
Ages	65 years and older (older adult)
NCT number	NCT04476953
Status	Enrolling by invitation
Official title	COVID-FISETIN: A Phase 2 Placebo-Controlled Pilot Study in SARS-CoV-2 of Fisetin to Alleviate Dysfunction and Excessive Inflammatory Response in Hospitalized Adults
Study start date	August 3, 2020
Study end date	September 2023 (final data collection date for primary outcome measure)
Study description	The purpose of this study is to test whether fisetin, a senolytic drug, can assist in preventing an increase in the disease's progression and alleviate complications of coronavirus due to an excessive inflammatory reaction To determine if fisetin treatment can prevent deterioration of oxygenation status as measured by S/F ratio: SpO_2/FiO_2, as well as prevent deterioration in physical function (frailty) and hyper-inflammation, other measures of oxygenation status (progression to supplemental oxygen requirement, assisted breathing/ -ventilation), and progression from mild/moderate to severely/critically proven SARS-CoV-2 infection in hospitalized patients and to evaluate the safety and tolerability of fisetin in this patient population
Condition	COVID-19

Study arms	• Experimental: Treatment group
	    Subjects will receive treatment drug (fisetin)
	      Intervention: Drug—Fisetin
	• Placebo comparator: Placebo group
	    Subjects will receive placebo
	      Intervention: Drug—Placebo

Study characteristic	Description
Estimated enrollment	80
Sex/gender	All sexes eligible for study
Ages	18 years and older (adult, older adult)
NCT number	NCT05482672
Status	Not yet recruiting
Official title	GetHealthy-OA: A Biopsychosocial Treatment Approach to Improve Pain and Function for Patients with Comorbid Knee Osteoarthritis, Obesity, and Depression
Study start date	July 1, 2023
Study end date	June 30, 2027 (final data collection date for primary outcome measure)
Study description	The investigators have previously identified knee osteoarthritis patients with the combination of depression and an unhealthy weight may be an increased risk of more rapid joint degeneration and worsening pain. The GetHealthy-OA program combines a mind-body program with the oral supplement fisetin to potentially reduce the risk for this population by treating psychosocial, mechanical, and inflammatory mechanisms of knee osteoarthritis. This randomized clinical trial will compare the GetHealthy-OA program to minimally enhanced usual care plus an oral placebo The study will employ a double-blind, randomized, placebo-controlled clinical trial to compare the multimodal GetHealthy-OA program versus placebo. As part of this clinical trial, 120 patients will be randomized to one of two groups: The GetHealthy-OA program that combines a mind-body program with oral fisetin versus a control group treated with an oral placebo and minimally enhanced usual care (MEUC). The GetHealthy-OA group will participate in the 6-week mind-body program and will take oral fisetin for two consecutive days (days 1 and 2), a 28-day washout period, and then another 2-day course (days 31 and 32). The MEUC group will be given a health education handout at the date of baseline testing and will take an oral placebo for two consecutive days (days 1 and 2) and then again 28 days later (days 29 and 30). To determine if immediate improvements are realized and whether gains are sustained after completing the program, participants will be assessed at baseline, 6 weeks, 3 months, and 6 months The GetHealthy-OA mind-body program is a live video, group program delivered via secure telehealth with mind-body skills to reduce pain and increase physical activity to promote optimal joint loading. The online program will be delivered remotely by a psychologist based at Massachusetts General Hospital via Zoom and we will assess symptoms and monitor any technical difficulties. The program consist of six sessions that will last about 45 min. The sessions are done on your smartphone or computer and will include a group of four to five other people with knee arthritis who are also taking part in the study with you at the same time. The GetHealthy-OA group will participate in the 6-week mind-body program and will take oral fisetin for two consecutive days (days 1 and 2), a 28-day washout period, and then another 2-day course (days 31 and 32)

Study characteristic	Description
	The MEUC group will be given a health education booklet at the date of baseline testing and will take an oral placebo (cornstarch) for two consecutive days (days 1 and 2) and then again 28 days later (days 29 and 30). The booklet will contain brief summarized information that reflects the active intervention topics including the trajectory of pain and recovery for those with knee osteoarthritis, the role of relaxation strategies to manage pain, and the importance of returning to engagement in activities of daily living. The MEUC will be given an oral placebo (cornstarch) that is identical to the fisetin capsules with the same dosing regimen as the GetHealthy-OA group. Similar to participants in GetHealthy-OA group, participants in this group will receive usual medical care as determined by the medical team. Usual care involves meetings with physicians, medical staff, and physical therapy. Usual care is identical in intervention and control groups.
Condition	• Knee osteoarthritis • Obese • Depression

Study arms	• Experimental: GetHealthy-OA
	    The GetHealthy-OA program combines a 6-week mind-body program delivered via live video with the oral supplement fisetin. Oral fisetin will be taken for two consecutive days (days 1 and 2), a 28-day washout period, and then another 2-day course (days 31 and 32)
	    Interventions:
	    • Drug: Fisetin
	    • Behavioral: GetHealthy-OA mind-body program
	• Placebo comparator: Minimally enhanced usual care
	    The minimally enhanced usual care group will be given a health education booklet at the date of baseline testing and will take an oral placebo for two consecutive days (days 1 and 2) and then again 28 days later (days 29 and 30)
	    Interventions:
	    • Drug: Placebo oral capsule
	    • Behavioral: Health education booklet

Estimated enrollment	120
Sex/gender	All sexes eligible for study
Ages	40 years and older (adult, older adult)
NCT number	NCT05653258
Status	Not yet recruiting
Official title	Single Nuclei RNA-Sequencing to Map Adipose Cellular Populations and Senescent Cells in Older Subjects
Study start date	August 2023
Study end date	February 2026 (final data collection date for primary outcome measure)

Study characteristic	Description
Study description	All participants will undergo baseline biopsies of subcutaneous abdominal adipose tissue for cellular/molecular profiling via snRNA-seq and metabolic/physiological assessments (insulin sensitivity, glucose tolerance, and β-cell function). Older obese participants will be randomized into three arms: lifestyle intervention (n = 24), senolytics (n = 24), or placebo (n = 24)
	All participants after consent and enrollment will undergo adipose tissue single nuclei RNA sequencing and metabolic phenotyping. Subjects will undergo glucose tolerance testing to document glucose tolerance status. Each subject will be provided an accelerometer to be worn on their dominant wrist for 7 days for assessment of habitual activity
	A dietitian will teach them to utilize the SmartIntake3 smartphone food picture application (app) for a 7-day food record. The app will be used to record the amount of each meal consumed in order to determine daily food and beverage and supplement intake and quantity for dietary composition analysis. DEXA analysis will be performed to measure lean and fat body mass
	Subjects will undergo evaluation of physical function/performance, including the Short Physical Performance Battery (SPPB) and VO2 peak testing for assessment of aerobic capacity. The SPPB will be done in older adults only
	The NIH Patient-Reported Outcomes Measurement System (PROMIS) will be used to measure participants' self-report of symptoms, function, and health-related quality of life in the domains of physical, mental, and social health
	All subjects will undergo a two-step euglycemic insulin clamp and indirect calorimetry
	Only older obese participants will continue to the randomization to lifestyle intervention, senolytic agents, or placebo
Condition	• Obesity • Healthy lifestyle

Study arms	• Younger lean group
	     Participants will be aged 18–30 years and have a BMI of 18.5–24.9 kg/m^2
	         Intervention: Procedure—Abdominal adipose tissue biopsy
	• Older lean group
	     Participants will be over 65 years of age with a BMI of 18.5–24.9 kg/m^2
	         Intervention: Procedure—Abdominal adipose tissue biopsy
	• Experimental: Older obese group
	     Participants will be over 65 years of age with a BMI of 30–39.9 kg/m^2
	         Interventions:
	         • Other: Lifestyle intervention
	         • Drug: Dasatinib 100 mg
	         • Drug: Quercetin 1000 mg
	         • Drug: Placebo
	         • Procedure: Abdominal adipose tissue biopsy

Estimated enrollment	120
Sex/gender	All sexes eligible for study
Ages	16 years and older (child, adult, older adult)
NCT number	NCT03675724
Status	Recruiting

Study characteristic	Description
Official title	AFFIRM-LITE: A Phase 2 Randomized, Placebo-Controlled Study of Alleviation by Fisetin of Frailty, Inflammation, and Related Measures in Older Adults
Study start date	November 15, 2018
Study end date	June 2024 (final data collection date for primary outcome measure)
Study description	This is a pilot study to test the efficacy of the anti-inflammatory drug (fisetin) in reducing inflammatory factors in blood in elderly adults and to test the efficacy of the drug (fisetin) in reducing frailty and markers of inflammation, insulin resistance, and bone resorption in elderly adults To the researchers' knowledge, there are no published studies utilizing fisetin in alteration of frailty markers. Several studies involve use of fisetin for its anti-oxidative and anti-apoptotic effects in animal models. Fisetin may reduce oxidative stress, alleviate hyperglycemia, and improve kidney function. No one has evaluated the biologic markers of inflammation and frailty in older adults. The researchers plan to evaluate markers of frailty and markers of inflammation, insulin resistance, and bone resorption while maintaining bone formation in older adults
Condition	Frail elderly syndrome

Study arms	• Experimental: Treatment
	   Fisetin 20 mg/kg/day, orally for two consecutive days
	     Intervention: Dietary supplement—Fisetin
	• Placebo comparator—Placebo
	   Placebo capsules orally for two consecutive days
	     Intervention: Drug—Placebo oral capsule

Estimated enrollment	40
Sex/gender	All sexes eligible for study
Ages	70 years and older (older adult)
NCT number	NCT03430037
Status	Recruiting
Official title	AFFIRM: A Phase 2 Randomized, Placebo-Controlled Study of Alleviation by Fisetin of Frailty, Inflammation, and Related Measures in Older Women
Study start date	February 6, 2018
Study end date	June 2024 (final data collection date for primary outcome measure)
Study description	This is a pilot study to evaluate whether targeting inflammation will help reduce markers of insulin resistance inflammation, bone resorption, and physical dysfunction in elderly women with gait disturbance. Positive results of this study would lead to the development of a larger clinical trial examining the effects of this intervention on age-related dysfunction To the researchers' knowledge, there are no published studies utilizing fisetin in alteration of frailty markers. Several studies involve use of fisetin for its anti-oxidative and anti-apoptotic effects in animal models. Fisetin may reduce oxidative stress, alleviate hyperglycemia, and improve kidney function. No one has evaluated the biologic markers of inflammation and frailty in older postmenopausal women. The researchers plan to evaluate markers of frailty and markers of inflammation, insulin resistance, and bone resorption while maintaining bone formation in older postmenopausal women

Study characteristic	Description
Condition	Frail elderly syndrome

Study arms	• Experimental: Treatment
	    Fisetin 20 mg/kg/day, orally for two consecutive days, for two consecutive months
	        Intervention: Dietary supplement—Fisetin
	• Placebo comparator: Placebo
	    Placebo capsules orally for two consecutive days, for two consecutive months
	        Intervention: Drug—Placebo oral capsule

Estimated enrollment	40
Sex/gender	Sexes eligible for study: Female Gender-based eligibility: Male
Ages	70 years and older (older adult)
NCT number	NCT04946383
Status	Active, not recruiting
Official title	The Safety and Effectiveness of Quercetin and Dasatinib on the Epigenetic Ageing Rates in Healthy Individuals
Study start date	December 16, 2020
Study end date	December 31, 2021 (final data collection date for primary outcome measure)
Study description	Assessing the effects of quercetin and dasatinib over a 16-week period on participants' epigenetic biological ageing. The patients are being tested at baseline, halfway point, and after the trial period This is a prospective non-randomized clinical study of 20–25 patients to evaluate the effectiveness of quercetin and dasatinib supplements on the patient's epigenetic ageing rate. The investigators predict that quercetin and dasatinib combined will slow cell proliferation and thus decelerate the rate of ageing There is evidence that quercetin and dasatinib slow cell proliferation and decelerate ageing and the risk of age-related diseases. The aim of this pilot study is to evaluate the safety, efficacy, and feasibility of quercetin and dasatinib as effective treatment options to improve clinical care of healthy individual's epigenetic ageing rate, thus prolonging longevity Despite considerable effort, successful treatment of reversing one's biological age has been shown to be a difficult therapeutic challenge. There is evidence that dasatinib + quercetin (quercetin and dasatinib) is a safe and effective treatment option to improve clinical care of healthy individual's biological age. Studies have shown that dasatinib + quercetin slows cell proliferation and decelerates ageing and the risk of age-related diseases
Condition	Ageing
Study arms	Experimental: Quercetin and dasatinib supplements 500 mg quercetin and 50 mg dasatinib oral capsules on Monday, Tuesday, Wednesday (3 days in a row) per month for the duration of 6 months Intervention: Drug—Dasatinib plus quercetin
Estimated enrollment	25
Sex/gender	All sexes eligible for study
Ages	40 years and older (adult, older adult)
NCT number	NCT05422885
Status	Recruiting

Study characteristic	Description
Official title	Senolytics to Alleviate Mobility Issues and Neurological Impairment in Ageing
Study start date	May 20, 2022
Study end date	June 1, 2023 (final data collection date for primary outcome measure)
Study description	The purpose of this pilot study is to demonstrate the safety and feasibility of administering intermittent doses of dasatinib and quercetin (D + Q) in older adults at risk of Alzheimer's disease (AD). The study will evaluate whether giving D + Q may improve cerebral blood flow regulation, mobility, and cognition in older adults and thus may prevent progression to Alzheimer's disease The investigators will conduct a 12-week single-arm, open-label, pre-post pilot study in 12 adults aged 65 or older with slow gait speed (<1.0 m/s) and mild cognitive impairment (MCI, defined as a Telephone Montreal Cognitive Assessment Score (MoCA) <19). Participants will be asked to take 100 mg of dasatinib and 1250 mg of quercetin for two consecutive days, every 2 weeks over a period of 12 weeks (12 doses in total, given over six cycles) At baseline, enrolled participants will undergo gait speed and neurocognitive testing and provide blood and urine to evaluate biomarkers of senescence. At visits 3, 4, 6, and 7, participants will have safety labs drawn, and the study team will assess medication adherence and adverse events. At visits 2, 5, and 8, participants will undergo cognitive assessments, gait speed testing, cerebral blood flow, and neurovascular coupling testing. At the final study visit, participants will again provide blood and urine to assess biomarkers of senescence
Condition	Ageing

Study arms	Experimental: Arm 1
	Dasatinib and quercetin
	Interventions:
	  • Drug: Dasatinib
	  • Drug: Quercetin

Estimated enrollment	12
Sex/gender	All sexes eligible for study
Ages	65 years and older (older adult)
NCT number	NCT04785300
Status	Enrolling by invitation
Official title	ALSENLITE: An Open-Label Pilot Study of Senolytics for Alzheimer's Disease
Study start date	July 6, 2022
Study end date	December 2023 (final data collection date for primary outcome measure)
Study description	This study is being done to evaluate the safety and feasibility of using dasatinib and quercetin together in subjects with mild cognitive impairment (MCI) or Alzheimer's disease The underlying processes driving chronic neurodegeneration in Alzheimer's disease (AD) and related neurodegenerative disorders are largely unknown. Ageing is the major risk factor for AD. Moreover, individuals with AD suffer from significantly more comorbid conditions than demographically matched older adults. This study is an open-label pilot study of intermittent administration of the senolytic drug regimen dasatinib (D) + quercetin (Q) in symptomatic adults over 55 with clinical diagnosis of probable Alzheimer's disease and Alzheimer's biomarker positivity by tau-PET

Study characteristic	Description
Condition	• Mild cognitive impairment • Alzheimer's disease

Study arms	Experimental: Dasatinib plus quercetin treatment group
	Subjects with MCI or Alzheimer's disease will take dasatinib and quercetin by mouth at the same times for 2 days out of every 15 days for six cycles lasting for a total of 77 days (12 concurrent doses of each agent)
	Interventions:
	  • Drug: Dasatinib
	  • Drug: Quercetin

Estimated enrollment	20
Sex/gender	All sexes eligible for study
Ages	55 years and older (adult, older adult)
NCT number	NCT02874989
Status	Completed
Official title	Targeted Removal of Pro-inflammatory Cells: An Open-Label Human Pilot Study in Idiopathic Pulmonary Fibrosis
Study start date	December 16, 2016
Study end date	June 3, 2019 (final data collection date for primary outcome measure)
Study description	The study team hypothesizes that intermittent (three doses administered over three consecutive days in three consecutive weeks) oral administration of combination dasatinib (100 mg/day) + quercetin (1250 mg/day) will be safe and well tolerated in patients with IPF. Treatment with D + Q will result in reduced abundance of pro-inflammatory cells within subjects over baseline. Finally, the reduction in biomarkers of cellular pro-inflammatory state will be related to no change in functional and patient-reported outcomes
Condition	Idiopathic pulmonary fibrosis (IPF)

Study arms	• Experimental: Dasatinib + quercetin
	  Intervention: Drug—Dasatinib + quercetin
	• Placebo comparator: Placebo
	  Intervention: Drug—Placebo

Actual enrollment	26
Sex/gender	All sexes eligible for study
Ages	50 years and older (adult, older adult)
NCT number	NCT02652052
Status	Recruiting
Official title	The investigators hope to find the proof-of-principle concept from this pilot study so that the investigators can design a clinical trial based on the results of the explanatory hypothesis
Study start date	March 1, 2016
Study end date	October 15, 2023 (final data collection date for primary outcome measure)

Study characteristic	Description
Study description	HSCT survivors are at an increased risk for premature ageing. No one has evaluated the biologic markers of premature ageing and senescence in HSCT survivors and their correlation with clinical outcomes, lifestyle, and nutrition. The investigators will evaluate age-related changes in HSCT survivors, with specified measures of premature ageing, and employ therapeutic opportunities based on targeting senescent cells by conducting the first in-human pilot study of senolytic drugs (in HSCT survivors utilizing a combination of senolytics)
Condition	Stem cell transplant

Study arms	• Group 1: Observational
	  Standard of care: Observation only
	    Intervention: Other—Standard of care (observation only)
	• Experimental: Group 2—Dasatinib and quercetin
	  Interventional: The drugs dasatinib and quercetin will be used in this arm
	  Interventions:
	  • Drug: Group 2—Quercetin
	  • Drug: Group 2—Dasatinib

Estimated enrollment	10
Sex/gender	All sexes eligible for study
Ages	18 years and older (adult, older adult)
NCT number	NCT04129944
Status	Completed
Official title	A Phase 2, Randomized, Double-Blind, Placebo-Controlled, Single-Dose Study of UBX0101 in Moderate to Severe, Painful Osteoarthritis of the Knee
Study start date	October 30, 2019
Study end date	May 20, 2020 (final data collection date for primary outcome measure)
Study description	A study to assess efficacy, safety, and tolerability of a single-dose intra-articular administration of UBX0101 in patients with moderate to severe painful knee osteoarthritis (OA) This is a randomized, double-blind, placebo-controlled, single-dose, parallel-group study to assess the efficacy, safety, and tolerability of a single-dose intra-articular (IA) administration of UBX0101 in patients with moderate to severe painful knee osteoarthritis (OA) Approximately 180 patients will be randomized (1:1:1:1) to one of four treatment groups (three dose levels of UBX0101 and placebo; approximately 45 patients per group), all administered by IA route at week 0. The four treatment groups will be enrolled concurrently The primary objective of the study is to evaluate the effect of IA administration of UBX0101 on the change from baseline to week 12 of pain in the target knee
Condition	• Osteoarthritis • Knee

Study arms	• Placebo comparator: Placebo
	  Intervention: Other—Placebo
	• Experimental: UBX0101 0.5 mg
	  Intervention: Drug—UBX0101
	• Experimental: UBX0101 2.0 mg
	  Intervention: Drug—UBX0101
	• Experimental: UBX0101 4.0 mg
	  Intervention: Drug—UBX0101

Study characteristic	Description
Actual enrollment	183
Sex/gender	All sexes eligible for study
Ages	40–85 years (adult, older adult)
NCT number	NCT04537884
Status	Completed
Official title	A Phase 1, Open-Label, Single Ascending Dose Study to Assess the Safety and Tolerability of a Single Intravitreal Injection of UBX1325 in Patients with Diabetic Macular Edema or Neovascular Age-Related Macular Degeneration
Actual start date	October 8, 2020
Actual end date	January 25, 2022 (final data collection date for primary outcome measure)
Study description	A study to evaluate safety, tolerability, and pharmacokinetics of a single intravitreal injection of UBX1325 in patients diagnosed with diabetic macular edema (DME) or neovascular age-related macular degeneration This is a phase 1, open-label, first-in-human (FIH), single-ascending dose (SAD) study consisting of approximately four cohorts. The total number of patients will be at least three per cohort plus three additional patients in the maximum tolerated dose (MTD) cohort in order to address the primary objective. Up to an additional 6 nAMD, patients will be enrolled in the highest dose cohort. A single dose of UBX1325 will be administered intravitreally and all patients will be followed for approximately 6 months
Condition	• Diabetic macular edema • Neovascular age-related macular degeneration
Study arms	Experimental: Treatment with UBX1325 UBX1325, single intravitreal injection, ascending dose Intervention: Drug—UBX1325
Actual enrollment	19
Sex/gender	All sexes eligible for study
Ages	50 years and older (adult, older adult)
NCT number	NCT04907253
Status	Recruiting
Official title	September 1, 2023 (final data collection date for primary outcome measure)
Actual start date	June 4, 2021
Study end date	September 1, 2023 (final data collection date for primary outcome measure)
Study description	The purpose of this study is to test the anti-inflammatory and anti-senescence effects of quercetin during coronary artery bypass graft surgery After being informed about the study and the potential risk, all patients giving written informed consent will be randomized in a double-blind manner (participant and investigators) on 1:1 ratio to receive quercetin (500 mg twice daily) or placebo (twice daily) starting 2 days before a coronary artery bypass graft surgery and for the duration of their hospitalization but up to 10 days (i.e., up to 7 days post-surgery). Blood (5 ml) will be collected the first morning after recruitment (t-1), 24 h post-surgery (t1), day 4 post-surgery (t2), and day of hospital discharge for blood analyses. During the surgery, if a discarded segment of mammary artery is available, it will be collected for laboratory work. Health status will be assessed during the follow-up visit 8–12 weeks post-surgery

Study characteristic	Description
Condition	Coronary artery disease

Study arms	• Active comparator: Quercetin
	    Patients receiving 500 mg quercetin twice daily
	      Intervention: Drug—Quercetin
	• Placebo comparator: Placebo
	    Patients receiving placebo twice daily
	      Intervention: Drug—Placebo

Estimated enrollment	100
Sex/gender	All sexes eligible for study
Ages	18 years and older (adult, older adult)
NCT number	*NCT02874924*
Status	Completed
Official title	Effect of Mammalian Target of Rapamycin Inhibition and Other Metabolism Modulating Interventions on the Elderly: Immune, Cognitive, and Functional Consequences
Study start date	June 2016
Study end date	September 2018 (final data collection date for primary outcome measure)
Study description	The ability to mount an effective immune response declines with age, leaving the elderly increasingly susceptible to infectious diseases and cancer. Rapamycin, an FDA-approved drug to prevent transplant rejection, increases the lifespan and healthspan of mice and ameliorates age-related declines in immune responsiveness, cancer survival, and cognition in laboratory animals. Investigators are conducting a translational trial to test whether rapamycin also improves life functions in humans focusing on elderly persons (aged 70–95).
Condition	Ageing

Study arms	• Experimental: Rapamycin
	    Rapamycin 1 mg taken once daily for 8 weeks
	      Intervention: Drug—Rapamycin
	• Placebo comparator: Placebo
	    Placebo taken once daily for 8 weeks
	      Intervention: Drug—Placebo
	• Experimental: Rapamycin Alone—Cardiovascular Effects
	    No placebo control; rapamycin 1 mg once daily for 8 weeks
	      Intervention: Drug—Rapamycin

Estimated enrollment	34
Sex/gender	All sexes eligible for study
Ages	70–95 years (older adult)
NCT number	*NCT03103893*
Status	Completed
Official title	Novel Compositions for Treating or Preventing Dermal Disorders
Study start date	September 25, 2017
Study end date	November 30, 2017 (final data collection date for primary outcome measure)

Study characteristic	Description
Study description	Ageing of the skin is the most prominent feature of the ageing process, being caused by multiple factors such as intrinsic ageing process and UV light exposure Dermal atrophy, also called skin atrophy or atrophy, is a disorder manifesting thinning or depression of the skin due to reduction of underlying tissue. Dermal atrophy is a major clinical problem in the elderly population. Loss of dermal integrity leads to increased fragility of the skin and precludes the use of intravenous lines in many cases. Skin tears are a significant concern in elderly individuals directly related to dermal atrophy. Impairment in wound healing is an important clinical sequelae of reduced dermal integrity leading to an increase in the number of the infections and complications following injury. Seborrheic keratosis, which comprises focal areas of epidermal thickening, can occur, possibly representing a response to damage. It has been estimated that 100% of individuals over 50 years of age harbor at least one of these lesion. There is not treatment for dermal atrophy and seborrheic keratoses require excision if they become large enough to cause discomfort or distress Therefore, there is a need to develop novel compositions and methods for treating or preventing certain age-related dermal conditions Rapamycin is an FDA-approved drug that has been in clinical use for over 15 years. Systemic application of rapamycin has been a central part of immunosuppressive therapy for transplant patients in combination with other immunosuppressants. The safety record for systemic use of rapamycin is excellent and few side effects are associated with extended use
Condition	Dermal atrophy
Study arms	Experimental: Rapamycin Rapamycin Intervention: Drug—Rapamycin
Estimated enrollment	36
Sex/gender	All sexes eligible for study
Ages	40–100 years (adult, older adult)
NCT number	NCT04200911
Status	Active, not recruiting
Official title	Cognition, Age, and RaPamycin Effectiveness-DownregulatIon of thE mTor Pathway (CARPE DIEM)
Study start date	June 1, 2020
Study end date	January 13, 2022 (final data collection date for primary outcome measure)
Study description	Evaluation of central nervous system penetration of orally administered Rapamune (RAPA) in older adults with mild cognitive impairment (MCI) or early Alzheimer's disease (AD) and investigation of associated safety, tolerability, target engagement, cognition, and functional status as initial proof-of-concept study This study is an open-label pilot study of orally administered RAPA to measure its target engagement in cerebrospinal fluid (CSF) and blood and to establish the feasibility and safety of RAPA treatment in older adults with MCI and early-stage AD as initial proof of concept for a larger phase 2 clinical trial

Study characteristic	Description
Condition	• Cognitive impairment, mild • Alzheimer's disease
Study arms	Experimental: RAPA intervention Sirolimus 1 mg orally once a day for 8 weeks Intervention: Drug—Rapamune
Estimated enrollment	10
Sex/gender	All sexes eligible for study
Ages	55–85 years (adult, older adult)
NCT number	*NCT05506488*
Status	Recruiting
Official title	Dasatinib and Quercetin: A Combination of Senolytics to Treat Fibrotic Non-alcoholic Fatty Liver Disease—The TRUTH Study
Study start date	March 1, 2023
Study end date	October 2024 (final data collection date for primary outcome measure)
Study description	To examine the effect of dasatinib plus quercetin on liver fibrosis in individuals with biopsy-proven NAFLD with fibrosis by performing a double-blind randomized controlled proof-of-principle study Non-alcoholic fatty liver disease (NAFLD) is estimated to affect approximately 25–30% of the population in Western countries and is now the leading cause of chronic liver disease globally. NAFLD is a progressive liver disease, and approximately 30% of individuals progress from simple steatosis to non-alcoholic steatohepatitis (NASH), which can further progress to cirrhosis and hepatocellular carcinoma. In the Netherlands, it is estimated that 2.5 million people have NAFLD, and this number is thought to increase by 50% in the next 10 years driven by an increasing prevalence of obesity and type 2 diabetes and an ageing population. Independent of other cardiometabolic diseases, cardiovascular disease is the leading cause of death in individuals with NAFLD, followed by extrahepatic malignancies and liver-related complications. NAFLD results in sustained healthcare costs and economic losses and reduced health-related quality of life It is now widely accepted that liver fibrosis is a result of liver injury secondary to NAFLD and is a major predictor for liver-related and overall mortality in individuals with NAFLD. The process of fibrosis progression is not completely understood, and it can vary considerably from one individual to another. Several risk factors for fibrosis progression have been identified: age, hypertension, obesity and type 2 diabetes. To date, no treatment is available that proved to be successful to target hepatic fibrosis. The only therapeutic options currently available therefore are the control of the concomitant metabolic diseases in addition to diet and lifestyle changes. Unfortunately, this inevitably will lead to polypharmacy and thereby decreases treatment adherence and increases the risk of adverse events and interactions with other drugs

Study characteristic	Description
	Recently, cellular senescence has been put forward as a causal factor in the development and progression of NAFLD and NAFLD-related liver fibrosis. Cellular senescence is one of the hallmarks of ageing and is defined as a stable arrest of the cell cycle coupled to specific phenotypic changes. Senescent cells secrete a collection of proteins called the senescence-associated secretory phenotype (SASP). This pro-inflammatory secretome drives age-related tissue dysfunction. Interestingly, metabolic dysregulation is thought to favor cellular senescence in several tissues involved in the pathogenesis of NAFLD such as the liver, pancreas, and adipose tissue, further perpetuating metabolic dysregulation. Of interest, cellular senescence can be targeted using senolytics. The combination of dasatinib, which is an EMA-approved tyrosine kinase inhibitor, and the antioxidant quercetin, which is a flavonol present in many fruits and vegetables, successfully clears senescent cells. Recent work in humans and rodents has shown that tissue function, including liver metabolism, can be recovered by clearing senescent cells with senolytics Due the potential role of senescence in NAFLD-related fibrosis, dasatinib plus quercetin might thus be an interesting future therapeutic option to tackle NAFLD-related fibrosis. Based on the long-term safety profile of these treatments and the high unmet clinical need as there currently is no treatment for NAFLD, we aim to perform a double-blind randomized controlled proof-of-principle study in which patients with NAFLD-related liver fibrosis will be treated with dasatinib plus quercetin intermittently 3 days per week for 3 weeks, followed by a 4-week medication-free period. Subsequently, this treatment cycle will be repeated three times
Condition	• NAFLD • NASH with fibrosis • Liver fibrosis

Study arms	• Active comparator: Dasatinib plus quercetin
	  Day 0, 15 per arm, randomization; week 7, blood, fibroscan, ECG, questionnaires; week 14, blood, fibroscan, ECG, questionnaires; week 21, blood, fibroscan, ECG, questionnaires, liver biopsy; end of study
	    Intervention: Drug—Dasatinib (100 mg) + quercetin (1000 mg)
	• Placebo comparator: Placebo
	    Day 0, 15 per arm, randomization; week 7, blood, fibroscan, ECG, questionnaires; week 14, blood, fibroscan, ECG, questionnaires; week 21, blood, fibroscan, ECG, questionnaires, liver biopsy; end of study
	    Intervention: Other—Placebo

Estimated enrollment	30
Sex/gender	All sexes eligible for study
Ages	18–65 years (adult, older adult)
NCT number	NCT04488601
Status	Active, not recruiting
Official title	Participatory Evaluation (of) Ageing (with) Rapamycin (for) Longevity Study (PEARL): A Prospective, Double-Blind, Placebo-Controlled Trial for Rapamycin in Healthy Individuals Assessing Safety and Efficacy in Reducing Ageing Effects
Study start date	January 1, 2020

Study characteristic	Description
Study end date	December 2023 (final data collection date for primary outcome measure)
Study description	This is a randomized, placebo-controlled trial into the safety and efficacy in reducing clinical measures of ageing in an older adult population A randomized, double-blind, placebo-controlled trial assessing the effects of low and high doses of intermittent rapamycin on a weekly schedule. The researchers aim to establish a long-term safety profile and determine the long-term efficacy of rapamycin in reducing clinical ageing measures and biochemical and physiological endpoints associated with declining health and ageing in healthy older adults
Condition	Ageing

Study arms	• Experimental: Rapamycin 5
	    Rapamycin 5 mg/week
	        Intervention: Drug—Rapamycin
	• Experimental: Rapamycin 10
	    Rapamycin 10 mg/week
	        Intervention: Drug—Rapamycin
	• Placebo comparator: Placebo 1
	    Placebo once per week
	        Intervention: Drug—Placebo

Estimated enrollment	150
Sex/gender	All sexes eligible for study
Ages	50–85 years (adult, older adult)
NCT number	NCT05237687
Status	Not yet recruiting
Official title	The Role of Sirolimus in Preventing Functional Decline in Older Adults
Study start date	April 2023
Study end date	September 2023 (final data collection date for primary outcome measure)
Study description	Ageing is associated with progressive impairment of tissue and organ function, resulting in increased susceptibility to chronic disease, frailty, and disability. Currently, there are limited treatment options to alter this inevitable process. The proposed work has the potential to identify a new therapeutic intervention to decrease ageing-related degenerative processes Rapamycin or sirolimus is a macrocyclic immunosuppressive drug that inhibits the mammalian target of rapamycin (mTOR). The mammalian target of rapamycin (mTOR) pathway is part of phosphoinositide 3-kinase (PI3K)/protein kinase B (AKT)/mammalian target of rapamycin (mTOR)-dependent pathway which is fundamentally linked to cell metabolism, proliferation, differentiation, and survival. This pathway is altered in a variety of diseases, including cancers, immunosuppressed states, and fibroproliferative diseases. The mTOR kinase is considered one of the leading regulators of this pathway. Changes in mTOR signaling are closely associated with inflammation, cell growth, and survival, leading to the development of chronic diseases. Recent evidence also suggests that mTOR inhibitors are promising modulators of the ageing process by slowing the mechanisms of ageing at the cellular level. There is a growing appreciation of the potential impact of sirolimus in slowing ageing processes and in prolonging healthy lifespan

Study characteristic	Description
	The proposed study addresses critical gaps in our understanding of the safety and efficacy of sirolimus in delaying ageing processes and is based on findings in animal studies and incidental clinical observations. The investigators will overcome potential biases with a randomized controlled trial. The proposed intervention study is intended to improve our insight into clinical outcomes leading to prevention of chronic diseases such as skin cancer and mortality. Our overarching hypothesis is that sirolimus is one of the first pharmacological agents that will impact the ageing process and chronic disease development. Specifically, the investigators aim to investigate whether sirolimus can reduce the occurrence or increase in biomarkers of ageing processes.
Condition	Ageing

Study arms	• Active comparator: Intervention
	    Patients randomized to the intervention will initially take 0.5 mg sirolimus. The dose will be adjusted weekly to obtain a sirolimus levels of 5–7 ng/ml whole blood in the first months. After the first month, the patient will have monthly blood work and will be followed in the clinic every 3 months. Functional assessment and ageing biomarkers will be obtained at baseline and 1-year follow-up. Completion of the 1-year treatment period will be followed by a follow-up visit 4 weeks later
	    Intervention: Drug—Sirolimus
	• No intervention: Control
	    Interventions: Standard of care; patients are not going to receive any additional intervention

Estimated enrollment	14
Sex/gender	All sexes eligible for study
Ages	55 years and older (adult, older adult)
NCT number	*NCT04742777*
Status	Recruiting
Official title	Effect of mTOR Inhibition and Other Metabolism Modulating Interventions on the Elderly: Immune, Cognitive, and Functional Consequences ((Substudy E-RAPA cMRI with LGE)
Study start date	February 1, 2022
Study end date	December 2023 (final data collection date for primary outcome measure)
Study description	The ability to mount an effective immune response declines with age, leaving the elderly increasingly susceptible to infectious diseases and cancer. Rapamycin, an FDA-approved drug to prevent transplant rejection, increases the lifespan and healthspan of mice and ameliorates age-related declines in immune responsiveness, cancer survival, and cognition in laboratory animals. Investigators are conducting a translational trial to test whether rapamycin also improves life functions in humans focusing on elderly persons (aged 70–95) Substudy E will evaluate the rapamycin and cardiac function
Condition	Ageing
Study arms	Experimental: Rapamycin Rapamycin 1 mg for 8 weeks Intervention: Drug—Rapamycin

Study characteristic	Description
Estimated enrollment	12
Sex/gender	Sexes eligible for study: Male
Ages	70–95 years (older adult)
NCT number	*NCT04629495*
Status	Recruiting
Official title	Rapamycin: Effects on Alzheimer's and Cognitive Health (REACH)
Study start date	August 11, 2021
Study end date	December 2023 (final data collection date for primary outcome measure)
Study description	This study will evaluate the safety, tolerability, and feasibility of 12-month oral rapamycin treatment in older adults with amnestic mild cognitive impairment (aMCI) and early-stage Alzheimer's disease (AD) The study will consist of a screening/baseline period of up to 90 days pre-study drug, with a 12-month (+3 day) treatment period with rapamycin, followed by a post-treatment assessment completed within 14 days of the final study drug dose, and a final assessment conducted 6 months (+14 days) after the final study drug dose. The study duration is not expected to exceed 90 weeks for participants
Condition	• Mild cognitive impairment • Alzheimer's disease

Study arms	• Active comparator: RAPA (rapamycin) treatment group
	    Subjects will receive active drug
	      Intervention: Drug—Rapamycin
	• Placebo comparator: Placebo group
	    Subjects will receive placebo
	      Intervention: Other—Placebo

Estimated enrollment	40
Sex/gender	All sexes eligible for study
Ages	55–89 years (adult, older adult)

Glossary

Age-associated diseases Age-associated diseases are illnesses that typically occur more frequently as biological age increases, many of which are caused by increased levels of cellular senescence (Borghesan et al., 2020).

Ageing Ageing is defined as the gradual accumulation of molecular and cellular damage over time, leading to a decline in physical and mental capacity, and an increased risk of disease and death (Borghesan et al., 2020).

Apoptosis Apoptosis is a natural and controlled process of cell death that regulates growth, development, maintaining tissue balance, and suppressing tumor growth (Borghesan et al., 2020; Pignolo et al., 2020).

Cellular senescence Cellular senescence is the process by which cells become permanently unable to divide, and do not undergo apoptosis. Senescent cells have several notable changes including the release of Senescence-Associated Secretory Phenotype (SASP). However, these changes vary depending on the cell type and the cause of senescence (Pignolo et al., 2020).

Geroscience hypothesis The geroscience hypothesis suggests that since biological ageing is the underlying cause of most chronic diseases and debilitating conditions, interventions that slow or reverse the process of biological age would also simultaneously prevent, delay, or alleviate multiple age-associated diseases (Pignolo et al., 2020).

Healthspan Healthspan is the period of an individual's life during which they are in good health and free from disabilities and diseases (Borghesan et al., 2020).

Immunosenescence Immunosenescence is the gradual decline in function of the immune system, particularly the adaptive immune system, because of ageing which leads to an increased risk of illness and death (Borghesan et al., 2020).

Lifespan Lifespan is the measure of the average amount of time a population lives from birth to death (Borghesan et al., 2020).

Senescence-Associated Secretory Phenotype (SASP) Senescence-Associated Secretory Phenotype (SASP) is chemicals secreted by senescent cells. In moderation transient levels of SASPs assist with wound healing and tumor suppression.

G. Bennett, *Senotherapy*, SpringerBriefs in Modern Perspectives on Disability Research, https://doi.org/10.1007/978-981-97-3637-9

However, heightened levels of SASPs can cause inflammation and senescent cell creation (Cuollo et al., 2020; Ohtani, 2022).

Senolytics Senolytics are medications that eliminate senescent cells (Kitaeva et al., 2024).

Senomorphics Unlike senolytics, which eliminate senescent cells entirely, senomorphics are medications that target specific functions of senescence cells, mainly SASP production and secretion, while keeping the cells alive (Di Micco et al., 2021).

References

Aatsinki, S. M., Buler, M., Salomäki, H., Koulu, M., Pavek, P., & Hakkola, J. (2014). Metformin induces PGC-1α expression and selectively affects hepatic PGC-1α functions. *British Journal of Pharmacology, 171*(9), 2351–2363. https://doi.org/10.1111/bph.12585

Algire, C., Moiseeva, O., Deschênes-Simard, X., Amrein, L., Petruccelli, L., Birman, E., Viollet, B., Ferbeyre, G., & Pollak, M. N. (2012). Metformin reduces endogenous reactive oxygen species and associated DNA damage. *Cancer Prevention Research (Philadelphia, Pa.), 5*(4), 536–543. https://doi.org/10.1158/1940-6207.CAPR-11-0536

Bauer, P. V., Duca, F. A., Waise, T. M. Z., Rasmussen, B. A., Abraham, M. A., Dranse, H. J., Puri, A., O'Brien, C. A., & Lam, T. K. T. (2018). Metformin alters upper small intestinal microbiota that impact a glucose-SGLT1-sensing glucoregulatory pathway. *Cell Metabolism, 27*(1), 101–117.e5. https://doi.org/10.1016/j.cmet.2017.09.019

Ben Sahra, I., Regazzetti, C., Robert, G., Laurent, K., Le Marchand-Brustel, Y., Auberger, P., Tanti, J. F., Giorgetti-Peraldi, S., & Bost, F. (2011). Metformin, independent of AMPK, induces mTOR inhibition and cell-cycle arrest through REDD1. *Cancer Research, 71*(13), 4366–4372. https://doi.org/10.1158/0008-5472.CAN-10-1769

Boni, J. P., Leister, C., Hug, B., Burns, J., & Sonnichsen, D. (2012). A single-dose placebo- and moxifloxacin-controlled study of the effects of temsirolimus on cardiac repolarization in healthy adults. *Cancer Chemotherapy and Pharmacology, 69*(6), 1433–1442. https://doi.org/10.1007/s00280-012-1845-7

Borghesan, M., Hoogaars, W. M. H., Varela-Eirin, M., Talma, N., & Demaria, M. (2020). A senescence-centric view of aging: Implications for longevity and disease. *Trends in Cell Biology, 30*(10), 777–791. https://doi.org/10.1016/j.tcb.2020.07.002

Bridgeman, S. C., Ellison, G. C., Melton, P. E., Newsholme, P., & Mamotte, C. D. S. (2018). Epigenetic effects of metformin: From molecular mechanisms to clinical implications. *Diabetes, Obesity & Metabolism, 20*(7), 1553–1562. https://doi.org/10.1111/dom.13262

Bruyn, G. A., Tate, G., Caeiro, F., Maldonado-Cocco, J., Westhovens, R., Tannenbaum, H., Bell, M., Forre, O., Bjorneboe, O., Tak, P. P., Abeywickrama, K. H., Bernhardt, P., van Riel, P. L., & RADD Study Group. (2008). Everolimus in patients with rheumatoid arthritis receiving concomitant methotrexate: A 3-month, double-blind, randomised, placebo-controlled, parallel-group, proof-of-concept study. *Annals of the Rheumatic Diseases, 67*(8), 1090–1095. https://doi.org/10.1136/ard.2007.078808

Cabreiro, F., Au, C., Leung, K. Y., Vergara-Irigaray, N., Cochemé, H. M., Noori, T., Weinkove, D., Schuster, E., Greene, N. D., & Gems, D. (2013). Metformin retards aging in C. elegans by altering microbial folate and methionine metabolism. *Cell, 153*(1), 228–239. https://doi.org/10.1016/j.cell.2013.02.035

G. Bennett, *Senotherapy*, SpringerBriefs in Modern Perspectives on Disability Research, https://doi.org/10.1007/978-981-97-3637-9

Cameron, A. R., Morrison, V. L., Levin, D., Mohan, M., Forteath, C., Beall, C., McNeilly, A. D., Balfour, D. J., Savinko, T., Wong, A. K., Viollet, B., Sakamoto, K., Fagerholm, S. C., Foretz, M., Lang, C. C., & Rena, G. (2016). Anti-inflammatory effects of metformin irrespective of diabetes status. *Circulation Research, 119*(5), 652–665. https://doi.org/10.1161/CIRCRESAHA.116.308445

Cheki, M., Shirazi, A., Mahmoudzadeh, A., Bazzaz, J. T., & Hosseinimehr, S. J. (2016). The radioprotective effect of metformin against cytotoxicity and genotoxicity induced by ionizing radiation in cultured human blood lymphocytes. *Mutation Research/Genetic Toxicology and Environmental Mutagenesis, 809*, 24–32. https://doi.org/10.1016/j.mrgentox.2016.09.001

Cheki, M., Ghasemi, M. S., Rezaei Rashnoudi, A., & Erfani Majd, N. (2021). Metformin attenuates cisplatin-induced genotoxicity and apoptosis in rat bone marrow cells. *Drug and Chemical Toxicology, 44*(4), 386–393. https://doi.org/10.1080/01480545.2019.1609024

Chen, J., Ou, Y., Li, Y., Hu, S., Shao, L. W., & Liu, Y. (2017a). Metformin extends *C. elegans* lifespan through lysosomal pathway. *eLife, 6*, e31268. https://doi.org/10.7554/eLife.31268

Chen, S. C., Brooks, R., Houskeeper, J., Bremner, S. K., Dunlop, J., Viollet, B., Logan, P. J., Salt, I. P., Ahmed, S. F., & Yarwood, S. J. (2017b). Metformin suppresses adipogenesis through both AMP-activated protein kinase (AMPK)-dependent and AMPK-independent mechanisms. *Molecular and Cellular Endocrinology, 440*, 57–68. https://doi.org/10.1016/j.mce.2016.11.011

Chung, M. M., Nicol, C. J., Cheng, Y. C., Lin, K. H., Chen, Y. L., Pei, D., Lin, C. H., Shih, Y. N., Yen, C. H., Chen, S. J., Huang, R. N., & Chiang, M. C. (2017). Metformin activation of AMPK suppresses AGE-induced inflammatory response in hNSCs. *Experimental Cell Research, 352*(1), 75–83. https://doi.org/10.1016/j.yexcr.2017.01.017

Chung, C. L., Lawrence, I., Hoffman, M., Elgindi, D., Nadhan, K., Potnis, M., Jin, A., Sershon, C., Binnebose, R., Lorenzini, A., & Sell, C. (2019). Topical rapamycin reduces markers of senescence and aging in human skin: An exploratory, prospective, randomized trial. *GeroScience, 41*(6), 861–869. https://doi.org/10.1007/s11357-019-00113-y

Cuollo, L., Antonangeli, F., Santoni, A., & Soriani, A. (2020). The Senescence-Associated Secretory Phenotype (SASP) in the challenging future of cancer therapy and age-related diseases. *Biology, 9*(12), 485. https://doi.org/10.3390/biology9120485

Cuyàs, E., Fernández-Arroyo, S., Verdura, S., García, R. Á., Stursa, J., Werner, L., Blanco-González, E., Montes-Bayón, M., Joven, J., Viollet, B., Neuzil, J., & Menendez, J. A. (2018a). Metformin regulates global DNA methylation via mitochondrial one-carbon metabolism. *Oncogene, 37*(7), 963–970. https://doi.org/10.1038/onc.2017.367

Cuyàs, E., Verdura, S., Llorach-Parés, L., Fernández-Arroyo, S., Joven, J., Martin-Castillo, B., Bosch-Barrera, J., Brunet, J., Nonell-Canals, A., Sanchez-Martinez, M., & Menendez, J. A. (2018b). Metformin is a direct SIRT1-activating compound: Computational modeling and experimental validation. *Frontiers in Endocrinology, 9*, 657. https://doi.org/10.3389/fendo.2018.00657

De Haes, W., Frooninckx, L., Van Assche, R., Smolders, A., Depuydt, G., Billen, J., Braeckman, B. P., Schoofs, L., & Temmerman, L. (2014). Metformin promotes lifespan through mitohormesis via the peroxiredoxin PRDX-2. *Proceedings of the National Academy of Sciences of the United States of America, 111*(24), E2501–E2509. https://doi.org/10.1073/pnas.1321776111

Di Micco, R., Krizhanovsky, V., Baker, D., & d'Adda di Fagagna, F. (2021). Cellular senescence in ageing: From mechanisms to therapeutic opportunities. *Nature Reviews. Molecular Cell Biology, 22*(2), 75–95. https://doi.org/10.1038/s41580-020-00314-w

Dickinson, J. M., Drummond, M. J., Fry, C. S., Gundermann, D. M., Walker, D. K., Timmerman, K. L., Volpi, E., & Rasmussen, B. B. (2013). Rapamycin does not affect post-absorptive protein metabolism in human skeletal muscle. *Metabolism: Clinical and Experimental, 62*(1), 144–151. https://doi.org/10.1016/j.metabol.2012.07.003

Diman, A., Boros, J., Poulain, F., Rodriguez, J., Purnelle, M., Episkopou, H., Bertrand, L., Francaux, M., Deldicque, L., & Decottignies, A. (2016). Nuclear respiratory factor 1 and endurance exercise promote human telomere transcription. *Science Advances, 2*(7), e1600031. https://doi.org/10.1126/sciadv.1600031

Dowling, R. J., Zakikhani, M., Fantus, I. G., Pollak, M., & Sonenberg, N. (2007). Metformin inhibits mammalian target of rapamycin-dependent translation initiation in breast cancer cells. *Cancer Research, 67*(22), 10804–10812. https://doi.org/10.1158/0008-5472.CAN-07-2310

Drummond, M. J., Fry, C. S., Glynn, E. L., Dreyer, H. C., Dhanani, S., Timmerman, K. L., Volpi, E., & Rasmussen, B. B. (2009). Rapamycin administration in humans blocks the contraction-induced increase in skeletal muscle protein synthesis. *The Journal of Physiology, 587*(Pt 7), 1535–1546. https://doi.org/10.1113/jphysiol.2008.163816

Dugel, P. U., Blumenkranz, M. S., Haller, J. A., Williams, G. A., Solley, W. A., Kleinman, D. M., & Naor, J. (2012). A randomized, dose-escalation study of subconjunctival and intravitreal injections of sirolimus in patients with diabetic macular edema. *Ophthalmology, 119*(1), 124–131. https://doi.org/10.1016/j.ophtha.2011.07.034

El-Mir, M. Y., Nogueira, V., Fontaine, E., Avéret, N., Rigoulet, M., & Leverve, X. (2000). Dimethylbiguanide inhibits cell respiration via an indirect effect targeted on the respiratory chain complex I. *The Journal of Biological Chemistry, 275*(1), 223–228. https://doi.org/10.1074/jbc.275.1.223

Fang, J., Yang, J., Wu, X., Zhang, G., Li, T., Wang, X., Zhang, H., Wang, C. C., Liu, G. H., & Wang, L. (2018). Metformin alleviates human cellular aging by upregulating the endoplasmic reticulum glutathione peroxidase 7. *Aging Cell, 17*(4), e12765. https://doi.org/10.1111/acel.12765

Fatt, M., Hsu, K., He, L., Wondisford, F., Miller, F. D., Kaplan, D. R., & Wang, J. (2015). Metformin acts on two different molecular pathways to enhance adult neural precursor proliferation/self-renewal and differentiation. *Stem Cell Reports, 5*(6), 988–995. https://doi.org/10.1016/j.stemcr.2015.10.014

Foretz, M., Hébrard, S., Leclerc, J., Zarrinpashneh, E., Soty, M., Mithieux, G., Sakamoto, K., Andreelli, F., & Viollet, B. (2010). Metformin inhibits hepatic gluconeogenesis in mice independently of the LKB1/AMPK pathway via a decrease in hepatic energy state. *The Journal of Clinical Investigation, 120*(7), 2355–2369. https://doi.org/10.1172/JCI40671

Gensler, G., Clemons, T. E., Domalpally, A., Danis, R. P., Blodi, B., Wells, J., 3rd, Rauser, M., Hoskins, J., Hubbard, G. B., Elman, M. J., Fish, G. E., Brucker, A., Margherio, A., & Chew, E. Y. (2018). Treatment of geographic atrophy with intravitreal sirolimus: The age-related eye disease study 2 ancillary study. *Ophthalmology Retina, 2*(5), 441–450. https://doi.org/10.1016/j.oret.2017.08.015

Gillespie, Z. E., Wang, C., Vadan, F., Yu, T. Y., Ausió, J., Kusalik, A., & Eskiw, C. H. (2019). Metformin induces the AP-1 transcription factor network in normal dermal fibroblasts. *Scientific Reports, 9*(1), 5369. https://doi.org/10.1038/s41598-019-41839-1

Gundermann, D. M., Walker, D. K., Reidy, P. T., Borack, M. S., Dickinson, J. M., Volpi, E., & Rasmussen, B. B. (2014). Activation of mTORC1 signaling and protein synthesis in human muscle following blood flow restriction exercise is inhibited by rapamycin. *American Journal of Physiology. Endocrinology and Metabolism, 306*(10), E1198–E1204. https://doi.org/10.1152/ajpendo.00600.2013

Hawley, S. A., Gadalla, A. E., Olsen, G. S., & Hardie, D. G. (2002). The antidiabetic drug metformin activates the AMP-activated protein kinase cascade via an adenine nucleotide-independent mechanism. *Diabetes, 51*(8), 2420–2425. https://doi.org/10.2337/diabetes.51.8.2420

Hörbelt, T., Kahl, A. L., Kolbe, F., Hetze, S., Wilde, B., Witzke, O., & Schedlowski, M. (2020). Dose-dependent acute effects of everolimus administration on immunological, neuroendocrine and psychological parameters in healthy men. *Clinical and Translational Science, 13*(6), 1251–1259. https://doi.org/10.1111/cts.12812

Howell, J. J., Hellberg, K., Turner, M., Talbott, G., Kolar, M. J., Ross, D. S., Hoxhaj, G., Saghatelian, A., Shaw, R. J., & Manning, B. D. (2017). Metformin inhibits hepatic mTORC1 signaling via dose-dependent mechanisms involving AMPK and the TSC complex. *Cell Metabolism, 25*(2), 463–471. https://doi.org/10.1016/j.cmet.2016.12.009

Kalender, A., Selvaraj, A., Kim, S. Y., Gulati, P., Brûlé, S., Viollet, B., Kemp, B. E., Bardeesy, N., Dennis, P., Schlager, J. J., Marette, A., Kozma, S. C., & Thomas, G. (2010). Metformin, independent of AMPK, inhibits mTORC1 in a rag GTPase-dependent manner. *Cell Metabolism, 11*(5), 390–401. https://doi.org/10.1016/j.cmet.2010.03.014

Kitaeva, K. V., Solovyeva, V. V., Blatt, N. L., & Rizvanov, A. A. (2024). Eternal youth: A comprehensive exploration of gene, cellular, and pharmacological anti-aging strategies. *International Journal of Molecular Sciences, 25*(1), 643. https://doi.org/10.3390/ijms25010643

Kraig, E., Linehan, L. A., Liang, H., Romo, T. Q., Liu, Q., Wu, Y., Benavides, A. D., Curiel, T. J., Javors, M. A., Musi, N., Chiodo, L., Koek, W., Gelfond, J. A. L., & Kellogg, D. L., Jr. (2018). A randomized control trial to establish the feasibility and safety of rapamycin treatment in an older human cohort: Immunological, physical performance, and cognitive effects. *Experimental Gerontology, 105*, 53–69. https://doi.org/10.1016/j.exger.2017.12.026

Krebs, M., Brunmair, B., Brehm, A., Artwohl, M., Szendroedi, J., Nowotny, P., Roth, E., Fürnsinn, C., Promintzer, M., Anderwald, C., Bischof, M., & Roden, M. (2007). The mammalian target of rapamycin pathway regulates nutrient-sensitive glucose uptake in man. *Diabetes, 56*(6), 1600–1607. https://doi.org/10.2337/db06-1016

Kulkarni, A. S., Gubbi, S., & Barzilai, N. (2020). Benefits of metformin in attenuating the hallmarks of aging. *Cell Metabolism, 32*(1), 15–30. https://doi.org/10.1016/j.cmet.2020.04.001

Lee, Y. S., Doonan, B. B., Wu, J. M., & Hsieh, T. C. (2016). Combined metformin and resveratrol confers protection against UVC-induced DNA damage in A549 lung cancer cells via modulation of cell cycle checkpoints and DNA repair. *Oncology Reports, 35*(6), 3735–3741. https://doi.org/10.3892/or.2016.4740

Lee, D. J. W., Hodzic Kuerec, A., & Maier, A. B. (2024). Targeting ageing with rapamycin and its derivatives in humans: A systematic review. *The Lancet. Healthy Longevity, 5*(2), e152–e162. https://doi.org/10.1016/S2666-7568(23)00258-1

Lu, M., Su, C., Qiao, C., Bian, Y., Ding, J., & Hu, G. (2016). Metformin prevents dopaminergic neuron death in MPTP/P-induced mouse model of Parkinson's disease via autophagy and mitochondrial ROS clearance. *International Journal of Neuropsychopharmacology, 19*(9), pyw047. https://doi.org/10.1093/ijnp/pyw047

Madiraju, A. K., Erion, D. M., Rahimi, Y., Zhang, X. M., Braddock, D. T., Albright, R. A., Prigaro, B. J., Wood, J. L., Bhanot, S., MacDonald, M. J., Jurczak, M. J., Camporez, J. P., Lee, H. Y., Cline, G. W., Samuel, V. T., Kibbey, R. G., & Shulman, G. I. (2014). Metformin suppresses gluconeogenesis by inhibiting mitochondrial glycerophosphate dehydrogenase. *Nature, 510*(7506), 542–546. https://doi.org/10.1038/nature13270

Mannick, J. B., Del Giudice, G., Lattanzi, M., Valiante, N. M., Praestgaard, J., Huang, B., Lonetto, M. A., Maecker, H. T., Kovarik, J., Carson, S., Glass, D. J., & Klickstein, L. B. (2014). mTOR inhibition improves immune function in the elderly. *Science Translational Medicine, 6*(268), 268ra179. https://doi.org/10.1126/scitranslmed.3009892

Mannick, J. B., Teo, G., Bernardo, P., Quinn, D., Russell, K., Klickstein, L., Marshall, W., & Shergill, S. (2021). Targeting the biology of ageing with mTOR inhibitors to improve immune function in older adults: Phase 2b and phase 3 randomised trials. *The Lancet. Healthy Longevity, 2*(5), e250–e262. https://doi.org/10.1016/S2666-7568(21)00062-3

Martin-Montalvo, A., Mercken, E. M., Mitchell, S. J., Palacios, H. H., Mote, P. L., Scheibye-Knudsen, M., Gomes, A. P., Ward, T. M., Minor, R. K., Blouin, M. J., Schwab, M., Pollak, M., Zhang, Y., Yu, Y., Becker, K. G., Bohr, V. A., Ingram, D. K., Sinclair, D. A., Wolf, N. S., Spindler, S. R., et al. (2013). Metformin improves healthspan and lifespan in mice. *Nature Communications, 4*, 2192. https://doi.org/10.1038/ncomms3192

Minturn, R. J., Bracha, P., Klein, M. J., Chhablani, J., Harless, A. M., & Maturi, R. K. (2021). Intravitreal sirolimus for persistent, exudative age-related macular degeneration: A pilot study. *International Journal of Retina and Vitreous, 7*, 1–10. https://doi.org/10.1186/s40942-021-00281-0

Moiseeva, O., Deschênes-Simard, X., St-Germain, E., Igelmann, S., Huot, G., Cadar, A. E., Bourdeau, V., Pollak, M. N., & Ferbeyre, G. (2013). Metformin inhibits the senescence-associated secretory phenotype by interfering with IKK/NF-κB activation. *Aging Cell, 12*(3), 489–498. https://doi.org/10.1111/acel.12075

Montvida, O., Shaw, J., Atherton, J. J., Stringer, F., & Paul, S. K. (2018). Long-term trends in anti-diabetes drug usage in the U.S.: Real-world evidence in patients newly diagnosed with type 2 diabetes. *Diabetes Care, 41*(1), 69–78. https://doi.org/10.2337/dc17-1414

Na, H. J., Park, J. S., Pyo, J. H., Lee, S. H., Jeon, H. J., Kim, Y. S., & Yoo, M. A. (2013). Mechanism of metformin: Inhibition of DNA damage and proliferative activity in Drosophila midgut stem cell. *Mechanisms of Ageing and Development, 134*(9), 381–390. https://doi.org/10.1016/j.mad.2013.07.003

Na, H. J., Pyo, J. H., Jeon, H. J., Park, J. S., Chung, H. Y., & Yoo, M. A. (2018). Deficiency of Atg6 impairs beneficial effect of metformin on intestinal stem cell aging in Drosophila. *Biochemical and Biophysical Research Communications, 498*(1), 18–24. https://doi.org/10.1016/j.bbrc.2018.02.191

Neumann, B., Baror, R., Zhao, C., Segel, M., Dietmann, S., Rawji, K. S., Foerster, S., McClain, C. R., Chalut, K., van Wijngaarden, P., & Franklin, R. J. M. (2019). Metformin restores CNS remyelination capacity by rejuvenating aged stem cells. *Cell Stem Cell, 25*(4), 473–485.e8. https://doi.org/10.1016/j.stem.2019.08.015

Noren Hooten, N., Martin-Montalvo, A., Dluzen, D. F., Zhang, Y., Bernier, M., Zonderman, A. B., Becker, K. G., Gorospe, M., de Cabo, R., & Evans, M. K. (2016). Metformin-mediated increase in DICER1 regulates microRNA expression and cellular senescence. *Aging Cell, 15*(3), 572–581. https://doi.org/10.1111/acel.12469

Nussenblatt, R. B., Byrnes, G., Sen, H. N., Yeh, S., Faia, L., Meyerle, C., Wroblewski, K., Li, Z., Liu, B., Chew, E., Sherry, P. R., Friedman, P., Gill, F., & Ferris, F., 3rd. (2010). A randomized pilot study of systemic immunosuppression in the treatment of age-related macular degeneration with choroidal neovascularization. *Retina (Philadelphia, Pa.), 30*(10), 1579–1587. https://doi.org/10.1097/IAE.0b013e3181e7978e

Ohtani, N. (2022). The roles and mechanisms of senescence-associated secretory phenotype (SASP): Can it be controlled by senolysis? *Inflammation and Regeneration, 42*(1), 11. https://doi.org/10.1186/s41232-022-00197-8

Onken, B., & Driscoll, M. (2010). Metformin induces a dietary restriction-like state and the oxidative stress response to extend C. elegans healthspan via AMPK, LKB1, and SKN-1. *PLoS One, 5*(1), e8758. https://doi.org/10.1371/journal.pone.0008758

Palma, J. A., Martinez, J., Millar Vernetti, P., Ma, T., Perez, M. A., Zhong, J., Qian, Y., Dutta, S., Maina, K. N., Siddique, I., Bitan, G., Ades-Aron, B., Shepherd, T. M., Kang, U. J., & Kaufmann, H. (2022). mTOR inhibition with sirolimus in multiple system atrophy: A randomized, double-blind, placebo-controlled futility trial and 1-year biomarker longitudinal analysis. *Movement Disorders: Official Journal of the Movement Disorder Society, 37*(4), 778–789. https://doi.org/10.1002/mds.28923

Pavlidou, T., Marinkovic, M., Rosina, M., Fuoco, C., Vumbaca, S., Gargioli, C., Castagnoli, L., & Cesareni, G. (2019). Metformin delays satellite cell activation and maintains quiescence. *Stem Cells International, 2019*, 5980465. https://doi.org/10.1155/2019/5980465

Petrou, P. A., Cunningham, D., Shimel, K., Harrington, M., Hammel, K., Cukras, C. A., Ferris, F. L., Chew, E. Y., & Wong, W. T. (2014). Intravitreal sirolimus for the treatment of geographic atrophy: Results of a phase I/II clinical trial. *Investigative Ophthalmology & Visual Science, 56*(1), 330–338. https://doi.org/10.1167/iovs.14-15877

Pignolo, R. J., Passos, J. F., Khosla, S., Tchkonia, T., & Kirkland, J. L. (2020). Reducing senescent cell burden in aging and disease. *Trends in Molecular Medicine, 26*(7), 630–638. https://doi.org/10.1016/j.molmed.2020.03.005

Ruddy, R. M., Adams, K. V., & Morshead, C. M. (2019). Age- and sex-dependent effects of metformin on neural precursor cells and cognitive recovery in a model of neonatal stroke. *Science Advances, 5*(9), eaax1912. https://doi.org/10.1126/sciadv.aax1912

Sant'Anna, J. R., Yajima, J. P., Rosada, L. J., Franco, C. C., Prioli, A. J., Della-Rosa, V. A., Mathias, P. C., & Castro-Prado, M. A. (2013). Metformin's performance in in vitro and in vivo genetic toxicology studies. *Experimental Biology and Medicine (Maywood, N.J.), 238*(7), 803–810. https://doi.org/10.1177/1535370213480744

Sarfstein, R., Friedman, Y., Attias-Geva, Z., Fishman, A., Bruchim, I., & Werner, H. (2013). Metformin downregulates the insulin/IGF-I signaling pathway and inhibits different uterine serous carcinoma (USC) cells proliferation and migration in p53-dependent or -independent manners. *PLoS One, 8*(4), e61537. https://doi.org/10.1371/journal.pone.0061537

Seyfarth, H. J., Hammerschmidt, S., Halank, M., Neuhaus, P., & Wirtz, H. R. (2013). Everolimus in patients with severe pulmonary hypertension: A safety and efficacy pilot trial. *Pulmonary Circulation, 3*(3), 632–638. https://doi.org/10.1086/674311

Slack, C., Foley, A., & Partridge, L. (2012). Activation of AMPK by the putative dietary restriction mimetic metformin is insufficient to extend lifespan in Drosophila. *PLoS One, 7*(10), e47699. https://doi.org/10.1371/journal.pone.0047699

Śmieszek, A., Stręk, Z., Kornicka, K., Grzesiak, J., Weiss, C., & Marycz, K. (2017). Antioxidant and anti-senescence effect of metformin on mouse olfactory ensheathing cells (mOECs) may be associated with increased brain-derived neurotrophic factor levels—An ex vivo study. *International Journal of Molecular Sciences, 18*(4), 872. https://doi.org/10.3390/ijms18040872

Sun, L., Xie, C., Wang, G., Wu, Y., Wu, Q., Wang, X., Liu, J., Deng, Y., Xia, J., Chen, B., Zhang, S., Yun, C., Lian, G., Zhang, X., Zhang, H., Bisson, W. H., Shi, J., Gao, X., Ge, P., Liu, C., et al. (2018). Gut microbiota and intestinal FXR mediate the clinical benefits of metformin. *Nature Medicine, 24*(12), 1919–1929. https://doi.org/10.1038/s41591-018-0222-4

Tsai, H. H., Lai, H. Y., Chen, Y. C., Li, C. F., Huang, H. S., Liu, H. S., Tsai, Y. S., & Wang, J. M. (2017). Metformin promotes apoptosis in hepatocellular carcinoma through the CEBPD-induced autophagy pathway. *Oncotarget, 8*(8), 13832–13845. https://doi.org/10.18632/oncotarget.14640

Vasamsetti, S. B., Karnewar, S., Kanugula, A. K., Thatipalli, A. R., Kumar, J. M., & Kotamraju, S. (2015). Metformin inhibits monocyte-to-macrophage differentiation via AMPK-mediated inhibition of STAT3 activation: Potential role in atherosclerosis. *Diabetes, 64*(6), 2028–2041. https://doi.org/10.2337/db14-1225

Vazquez-Martin, A., Oliveras-Ferraros, C., Cufí, S., Martin-Castillo, B., & Menendez, J. A. (2011). Metformin activates an ataxia telangiectasia mutated (ATM)/Chk2-regulated DNA damage-like response. *Cell Cycle (Georgetown, Tex.), 10*(9), 1499–1501. https://doi.org/10.4161/cc.10.9.15423

Wang, G. Y., Bi, Y. G., Liu, X. D., Zhao, Y., Han, J. F., Wei, M., & Zhang, Q. Y. (2017). Autophagy was involved in the protective effect of metformin on hyperglycemia-induced cardiomyocyte apoptosis and Connexin43 downregulation in H9c2 cells. *International Journal of Medical Sciences, 14*(7), 698–704. https://doi.org/10.7150/ijms.19800

Wang, Y., Xu, W., Yan, Z., Zhao, W., Mi, J., Li, J., & Yan, H. (2018). Metformin induces autophagy and G0/G1 phase cell cycle arrest in myeloma by targeting the AMPK/mTORC1 and mTORC2 pathways. *Journal of Experimental & Clinical Cancer Research, 37*(1), 63. https://doi.org/10.1186/s13046-018-0731-5

Wang, Y., Liu, B., Yang, Y., Wang, Y., Zhao, Z., Miao, Z., & Zhu, J. (2019). Metformin exerts antidepressant effects by regulated DNA hydroxymethylation. *Epigenomics, 11*(6), 655–667. https://doi.org/10.2217/epi-2018-0187

Wen, H. Y., Wang, J., Zhang, S. X., Luo, J., Zhao, X. C., Zhang, C., et al. (2019). Low-dose sirolimus immunoregulation therapy in patients with active rheumatoid arthritis: A 24-week follow-up of the randomized, open-label, parallel-controlled trial. *Journal of Immunology Research, 2019*. https://doi.org/10.1155/2019/7684352

Wheaton, W. W., Weinberg, S. E., Hamanaka, R. B., Soberanes, S., Sullivan, L. B., Anso, E., Glasauer, A., Dufour, E., Mutlu, G. M., Budigner, G. S., & Chandel, N. S. (2014). Metformin inhibits mitochondrial complex I of cancer cells to reduce tumorigenesis. *eLife, 3*, e02242. https://doi.org/10.7554/eLife.02242

Wu, L., Zhou, B., Oshiro-Rapley, N., Li, M., Paulo, J. A., Webster, C. M., Mou, F., Kacergis, M. C., Talkowski, M. E., Carr, C. E., Gygi, S. P., Zheng, B., & Soukas, A. A. (2016). An ancient, unified mechanism for metformin growth inhibition in C. elegans and cancer. *Cell, 167*(7), 1705–1718.e13. https://doi.org/10.1016/j.cell.2016.11.055

Yan, Q., Han, C., Wang, G., Waddington, J. L., Zheng, L., & Zhen, X. (2017). Activation of AMPK/mTORC1-mediated autophagy by metformin reverses Clk1 deficiency-sensitized dopaminergic neuronal death. *Molecular Pharmacology, 92*(6), 640–652. https://doi.org/10.1124/mol.117.109512

Zhong, T., Men, Y., Lu, L., Geng, T., Zhou, J., Mitsuhashi, A., Shozu, M., Maihle, N. J., Carmichael, G. G., Taylor, H. S., & Huang, Y. (2017). Metformin alters DNA methylation genome-wide via the H19/SAHH axis. *Oncogene, 36*(17), 2345–2354. https://doi.org/10.1038/onc.2016.391